Nature's Blueprint

Also by Dan Hooper

*Dark Cosmos: In Search of Our Universe's
Missing Mass and Energy*

 Smithsonian Books

COLLINS
An Imprint of HarperCollins Publishers

Nature's Blueprint

*Supersymmetry and
the Search for a Unified
Theory of Matter and Force*

DAN HOOPER

All illustrations throughout the text are copyright © 2008 by
Patricia Wynne, except for:
Page 15, (image 001) © American Institute of Physics,
Emilio Segre Visual Archives.
Page 47, (image 005) Reprinted figure with permission from Carl D.
Anderson, *Physical Review*, vol. 43, p. 491 (1933). Copyright © 2004 by the
American Physical Society.
Page 69, (image 007) © American Institute of Physics, Emilio Segre
Visual Archives, Marshak Collection.
Page 109, (image 015) with permission from the Hebrew University of
Jerusalem.
Page 132, (image 020) with permission from Howard Georgi.
Page 183, (image 024) © Peter Reid and SCI-FUN.

FIRST EDITION

Designed by Nicola Ferguson

Library of Congress Cataloging-in-Publication Data is available upon request.

ISBN: 978-0-06-155836-8

08   09   10   11   12   OV/RRD   10   9   8   7   6   5   4   3   2   1

Contents

Acknowledgments

I would like to thank all of those who provided me with the suggestions, assistance, encouragment, and inspiration that made this book possible. In particular, I owe special thanks to Antony Harwood, T. J. Kelleher, Cheryl Potts, and Pasquale Serpico. Josh Friess, in contrast, contributed absolutely nothing to this project.

—Dan Hooper

Nature's Blueprint

> **The real voyage of
> discovery consists not in
> seeking new landscapes but in
> having new eyes.**
> —*Marcel Proust*

|||||||||| **1** ||||||||||

Discovery!

To the curious, nothing is more exciting than discovery. Nothing is more powerful, and nothing is more awe inspiring. There is something truly and profoundly irresistible about the act of learning secrets—in knowing what had once been hidden. Scientists, and science itself, are driven by this fascination. The secrets that science seeks are those belonging to nature. Nature presents us with the grandest of all puzzles. With every new advance or insight, we get a further glimpse into the inner workings of our world—the very blueprint of nature. Today, we are preparing to sneak a deeper and more detailed glimpse at this blueprint. We are poised upon the very edge of discovery.

For more than thirty years, physicists have been investigating a theory known as supersymmetry. Supersymmetry is a framework—a principle, really—that describes and explains the relationship between two of the most fundamental concepts in physics: matter and force. This is a theory possessing the highest degree of mathematical beauty and elegance. On a more practical note, the presence of supersymmetry also has the ability to solve many of the long-standing problems of particle physics. To date, however, experimental confirmation that supersymmetry actually exists has remained elusive. If or when evidence for supersymmetry is finally observed, it will be a monumental, era-defining moment in the history of science, on par with the greatest discoveries of Einstein, Newton, and Galileo. It will be the discovery of a lifetime.

The theory of supersymmetry predicts that many unseen kinds of matter must exist. This matter takes the form of particles called superpartners. Despite the efforts of many hundreds of physicists conducting experiments in search of these particles, no superpartners have ever been observed or detected. They remain hidden, at least for the time being. This has had little effect in deterring the theoretical physicists who passionately expect nature to be formulated in this way—to be supersymmetric. To many of these scientists, the ideas behind supersymmetry are simply too beautiful and too elegant not to be part of our universe. They solve too many problems and fit into our world too naturally. To these true believers, the superpartner particles simply must exist. That they remain hidden is merely a standing challenge to future physicists and the experiments they conduct.

Supersymmetry may not remain hidden from us for much longer. In fact, many of the world's most prominent physicists think it likely that the elusive superpartners are about to be revealed. To accomplish this feat, a remarkable machine has been built. This machine, called the Large Hadron Collider—the LHC for short—is an enormous particle accelerator located beneath and around the city of Geneva, Switzerland, and extending across the border into France. Through a circular underground tunnel, seventeen miles in circumference, protons will be accelerated by ultra-powerful magnets to amazing speeds—99.9999991 percent of the speed of light, or more than 670 million miles per hour. When two beams of protons are collided head-on, so much energy will be compressed into one place at one time that entirely new and unknown forms of matter will be able to be brought into existence. Among these new forms of matter, many believe, will be the superpartner particles predicted by the theory of supersymmetry. With this incredible machine, humankind will finally learn whether supersymmetry is—or is not—built into the very fabric of our universe.

BEAUTY CAN BE A difficult thing to understand, and an even more difficult thing to define. Although we all have some idea of what it means for something to be beautiful, it is very hard to put our finger on the appeal of a beautiful sound, image, or idea. Whether found in a da Vinci masterpiece, a Beethoven symphony, a Shakespeare sonnet, or a magnificent sunset, many of us feel that we can recognize beauty when we see or hear it. Ultimately, however, we also recognize that this is a subjective quality—something of ourselves that we project onto that which we perceive. Beauty truly is in the eye of the beholder.

This intrinsic subjectivity can even be found in the paragraph you just read. In it, I chose to use da Vinci, Beethoven, and Shakespeare as examples of artistic beauty because I somehow imagine that they might relate to most people, including those reading this book. Personally, I find far more beautiful the works of Pablo Picasso, the Rolling Stones, and John Steinbeck. You probably have other feelings of your own about beauty. Whether you find the work of Beethoven, the Rolling Stones, or Celine Dion more beautiful, you are neither right nor wrong.[1] Subjectivity is the essence of beauty.

---

[1] Unless you picked Celine Dion, in which case you are objectively wrong.

So if beauty is purely subjective, how can science, which strives to be objective, find something—such as the idea of supersymmetry—to be beautiful? Science has never produced an equation that could be used to calculate the quantity of beauty possessed by *anything*. Nor will it in the future. Beauty is not something that can be quantified. Perhaps sociologists could scientifically conduct surveys to learn how beautiful a subject appears to a given individual or set of individuals, but this tells us something only about the people being surveyed, and not about the subject itself. It is absolutely beyond the reach of science to judge beauty.

Although beauty may be beyond the purview of science, it is certainly not beyond the nature of scientists. Scientists are—beneath their lofty goals of objectivity—creatures of flesh and blood like everyone else. And like everyone else, we scientists see beauty in the world all around us. We may not be able to prove rigorously that one scientific idea is more beautiful than another, but we do experience and appreciate beauty. Like all human beings, we feel that we know beauty when we see it.

To the eyes of this beholder, the most profound forms of beauty are not those that can be found in the sounds or images of this world, but in those rare and special ideas of exceptional elegance and power. Take, for example, the maxim *Do unto others as you would have them do unto you*. This remarkable phrase contains within it a sense of judgment, universality, and balance. The same could be said of the Jeffersonian declaration that "all men are created equal"—although "all people are created equal" would have been even more beautiful, in my opinion. The Darwinian mechanism of natural selection has the remarkable power to bring forth the great diversity and complexity of life from primordial simplicity. The evolution of our universe from its simple and singular Big Bang origin to the rich and varied cosmos we witness today is similarly awe inspiring.

These are beautiful ideas.

Within science, what are often thought of as the most beautiful ideas are those able to explain a great many phenomena with only a simple concept or principle. Taken to the extreme limit, the uniquely most beautiful of all possible scientific theories would be a single idea from which could be derived absolutely every aspect of reality—a perfect theory of everything. The most beautiful theory could be written simply and briefly, perhaps even as a single equation. Although it need not be easy to do so, it would be possible to study that equation and with it answer any question about our world. As I am writing this paragraph, I can feel my heartbeat rise just a little and my palms begin to sweat. I sometimes react the same way to an exceptionally beautiful piece of music.

Sadly, we have no such perfect theory of everything. We do, however, have a number of scientific ideas that manage very concisely to explain a great deal about our world. Isaac Newton showed us that the force pulling objects toward Earth is the same that causes planets to move along their orbits. With this insight, Newton took what had been thought of as two unrelated aspects of our world and brought them together under the single concept of gravity. Before the middle of the nineteenth century, electricity and magnetism were thought to be entirely separate and unrelated occurrences. James Clerk Maxwell showed that they were merely different manifestations of the same phenomena. A magnetic field, it turns out, is nothing more than an electric field in motion. Shortly after the turn of the twentieth century, as part of his great theory of relativity, Albert Einstein taught us that the mass pulled upon by the force of gravity is the same mass that makes something difficult to move through its inertia.

These great developments—and others—in the history of science can each be described as moments of unification. Whenever two or more apparently separate phenomena are understood to be different facets of each other, or whenever they are shown to come from a common origin, we move one step closer to building a complete theory of everything. With each step, our understanding of our world becomes simpler, deeper, more elegant, and more beautiful.

Over the years and centuries, physicists have been remarkably successful in their efforts to unify the various aspects of their science. At one time, there were many separate and seemingly unrelated theories used to describe the countless types of observed physical phenomena—light, sound, heat, the motion of the planets, the force of gravity, electricity, magnetism, radioactivity, friction, and so on. Today, all of these aspects of our world—and many others—can be understood using only two theories. One of these is Einstein's theory of relativity. The other is the theory of quantum physics, or, more precisely, quantum field theory. The first of these theories describes space, time, and the force of gravity. The second describes the particles that make up the various types of matter in our world, along with the electromagnetic and nuclear forces. These two theories are, together, able to describe every single known phenomenon in our world.

ALTHOUGH FEW PHYSICISTS OFTEN do so, one can think of quantum field theory as two separate theories. Despite the fact that there is only one sequence of quantum physics courses offered at most universities, and only one quantum field theory textbook used by most students, there are effectively two very different quantum field theories—or at least two different sides of quantum field theory. These two quantum field theories describe two classifications of particles, known as fermions and bosons. Fermions are the particles that we normally think of as matter. The electrons, protons, and neutrons that make up all of the world's atoms and molecules are fermions, for example. In addition to these familiar particles, other more exotic varieties of fermions exist as well, and carry strange names such as neutrinos, muons, taus, and quarks. In the first version of the quantum theory to be formulated, fermions were all that existed. This theory described a world of matter and nothing else.

In the mid-1930s, the other side of quantum field theory was beginning to be understood for the first time. The equations behind this theory describe another kind of particle—bosons. Bosons are, in a sense, matter just like fermions, but they are also something more. Boson particles are the transmitters of force. Individual particles of light photons—are the bosons responsible for the transmission of the electromagnetic force. Simply put, photons *are* the electromagnetic force. Similarly, bosons called gluons constitute the force that holds together the nuclei of atoms—the strong force. Particles known simply as the W and Z bosons generate the force that brings forth certain types of radioactivity—the weak force. Without the presence of boson particles, there is no force. Bosons are force, and forces are the manifestations of bosons.

For decades following the birth of quantum field theory, fermions and bosons were thought of as separate aspects of the quantum nature of our world. It was understood how each behaves, and how they interact with each other, but ultimately fermions and bosons were thought of as simply different phenomena—phenomena that could have, in principle, existed without each other. Supersymmetry changes all of that. Before Isaac Newton, the behavior of orbiting planets was not known to have anything to do with the force that pulls objects toward Earth. Before James Clerk Maxwell, it was not known that magnetism was simply a manifestation of an electric field. Without supersymmetry, bosons and fermions are independent and unrelated aspects of our world. In the presence of supersymmetry, they are inseparable aspects of the same reality. Supersymmetry unifies the concepts of matter and force into a single theoretical framework—a framework in which fermions cannot exist without bosons, and bosons cannot exist without fermions.

OVER THE PAST THREE decades, the theory of supersymmetry has become something of an obsession for the worldwide community of particle physicists. Several tens of thousands of articles have been published in scientific journals on the subject. Several major conferences focus on it each year. Dozens of textbooks have been written on it. Considering that supersymmetry has yet to be experimentally confirmed to exist, this degree of study is remarkable— perhaps uniquely so. With the possible exception of string theory, I can think of no other unconfirmed idea that has ever been the focus of so much scientific research, or consumed so many scientific resources.

The amount of time and money devoted to the pursuit of supersymmetry is staggering. It is hard to find a particle physicist who has *not* worked on this theory at some time in his or her career. This obsession is not by any means a local phenomenon. I have seen scientific presentations on supersymmetry given in more than a dozen countries. All over the world, thousands of scientists have been imagining a beautifully supersymmetric universe. It is the dream of this multitude of physicists that soon we will finally discover the presence of supersymmetry in our world.

It is perhaps not surprising that supersymmetry is so fascinating to so many physicists. Matter and force are two of the most fundamental and broadly encompassing concepts in all of science, and have been studied for as long as human beings have been capable of pondering their world. Matter and force were each central to the science and philosophy of the ancient Greeks. Aristotle wrote about force being the cause of all motion, removing objects from their natural state of rest. While some of the ancient Greeks—such as

Leucippus, Democritus, and Epicurus—argued that matter was made up of indivisible pieces, called atoms, others—most notably the Stoics—insisted upon the continuous, and forever divisible, nature of matter.

As time passed and our understanding of the world changed, force and matter remained at the focus of science. By the Middle Ages, Aristotle's ideas about force had been challenged and improved upon by the Islamic philosopher Avicenna, and earlier by the neo-Platonist Johannes Philoponus. With the birth of the Renaissance and European Enlightenment, Galileo and Newton developed the ideas and equations describing force that are still taught in high school and university physics courses today.

Ideas regarding the nature of matter took somewhat longer to mature. Between the time of the ancient Greek philosophers and the European Enlightenment, the conception of matter had changed very little. It wasn't until the early nineteenth century that John Dalton introduced aspects of the modern atomic theory. Gradually, the periodic table of the elements was formulated and each of its members discovered. The atoms of this table, however, turned out not to be the atoms envisioned by Democritus and the other ancient Greeks. Whereas Democritus's atoms were absolutely indivisible, the periodic table contains only objects that can be further divided into even smaller parts. Atoms, after all, can be broken apart into protons, neutrons, and electrons. Protons and neutrons are furthermore made up of quarks, and quarks are held together by gluons. As far as we know, quarks, gluons, and electrons cannot be subdivided any further. They may indeed be examples of indivisible atoms, as once proposed by the philosophers of ancient Greece.

GIVEN THAT SUPERSYMMETRY IS of interest—perhaps obsessively so—to physicists all over the world, the story of its invention is worth telling. Just as the pursuit of supersymmetry is today a global effort, the very conception of this theory was a truly international endeavor.

The birth of supersymmetry came in the early 1970s, with the work of a handful of Soviet mathematical physicists. In 1971, Evgeny Likhtman and Yuri Golfand invented a mathematical theory in which you could take out all of the fermions, replace them with bosons, and simultaneously remove all of the bosons, replacing them with fermions, and get exactly the same thing that you started with. For the life of me, I can't understand why anyone would have wasted their time with such a crazy-sounding exercise. It is like imagining a world in which you could replace all of the trees with cars and vice versa without changing the way the world works. Even if you could do it, what would be the point? Well, Soviet scientists must have thought it was interesting because in 1972 another pair of them, Dmitry Volkov and Vladimir Akulov, invented another version of essentially the same theory.

This all took place at the height of the Cold War. The many thousands of Soviet and American nuclear weapons set on hairpin alert certainly did not help to encourage dialogue between the Soviet scientists and their counterparts in the west. Physicists such as Likhtman, Golfand, Akulov, and Volkov were almost never allowed to leave the Soviet Union, and very rarely published their results in non-Soviet journals. Although most of the Soviet journals were, in principle, available in the West—and in some cases even translated—they more often than not were unappreciated, poorly understood, and even unnoticed. Like many other important ideas during the Cold War, supersymmetry remained unknown in the West—but not for long. In 1973, two European physicists, Bruno Zumino and Julius Wess, independently developed the idea of supersymmetry, entirely unaware of the similar work by their Russian counterparts.

It was not immediately clear to many physicists what supersymmetry had to offer. It was not known to solve any problems, nor did it seem to be required for any other particularly compelling reason. It was mathematical physics in the purest sense—entirely theoretical, with little chance of having much to do with reality. Over the years since the invention of supersymmetry theory, however, this attitude has transformed dramatically.

I mentioned earlier that modern physics is built upon two great theories—quantum field theory and Einstein's general theory of relativity. Einstein's theory describes space, time, and the force of gravity spectacularly well. Despite its successes, however, we know that it is ultimately incompatible with our understanding of the quantum world. At very high temperatures, such as those in the first instants following the Big Bang, general relativity breaks down and no longer works. Under these extraordinary circumstances, another theory is needed to describe simultaneously the behavior of gravity and the role of quantum particles—a theory of quantum gravity. The task of building a theory of quantum gravity has been taken up by countless physicists, so far without success. Although there are promising avenues being pursued—string theory and loop quantum gravity, for example—no workable theory has yet materialized. Merging general relativity with quantum field theory is perhaps the greatest outstanding challenge of modern science.

By the mid-1970s, physicists began to appreciate that supersymmetry is not merely a relationship between fermions and bosons, but is also a symmetry related to space and time—related in a way that suggests that it could have something important to tell us about the nature of quantum gravity. Supersymmetry is able to expand the symmetries built into Einstein's theory of gravity in such a way that it might, it seems, alleviate some of the problems normally experienced when trying to combine general relativity with quantum field theory. In this way, supersymmetry can be promoted to something known as "supergravity." Since the mid-1980s, supersymmetry has also been understood to be an essential requirement of the leading candidate for a viable theory of quantum gravity—string theory, or superstring theory as it has since become known. Our best ideas of how gravity, space, and time fit into a quantum framework all point us toward supersymmetry.

Even if we put aside the issues with gravity and general relativity for the time being, there are other compelling reasons why so many particle physicists expect someday to discover supersymmetry. The three forces described by our quantum theory—the electromagnetic force and the strong and weak forces—and all of the known forms of matter fit together rather nicely—like well made puzzle pieces—in what are known as grand unified theories, or GUTs for short. Such theories describe all types of matter and all of the known forces—excluding gravity—in a remarkably simple and elegant fashion. Supersymmetry, physicists now understand, is an essential feature of these theories.

Although GUTs can be simple and elegant, without supersymmetry they are sick. Considering only the known forms of matter, the three quantum forces *almost* merge together into a single GUT force at very high energies, but not quite. If supersymmetry is included, however, they come together perfectly. Additionally, the mathematical structure behind supersymmetry manages to stabilize the masses of key particles, preventing them from obtaining and possessing catastrophic amounts of mass and energy.

As grand unified theories became increasingly popular among particle physicists in the late 1970s and early 1980s, so did supersymmetry. As it became understood in the mid-1980s that string theory requires supersymmetry, even more physicists became advocates. In the decades since its invention, the theory of supersymmetry has made a complete transition from mathematical abstraction—little more than a game or a clever trick—to what is now widely considered to be an absolutely integral part of modern physics. To its advocates, supersymmetry is an essential aspect of our universe's structure.

As the Large Hadron Collider prepares to begin its operation, many believe that this essential aspect of our universe's structure is about to be revealed.

THE MAIN PURPOSE OF this chapter is not to explain what supersymmetry is, or why so many physicists think it is likely to exist. Instead, it is simply to try to share with you how exciting it feels to be so close to witnessing this beautiful aspect of nature for the first time. Discovery is a truly amazing experience.

In the following chapters of this book, I will describe not only what supersymmetry is, but also how it fits into our world as we currently understand it—from its role in gravity and grand unification, to its place in the Big Bang and evolution of our universe. But before leaping right into supersymmetry, there are other symmetries built into the fabric of our world that I would like first to explore. Although, for the past several decades, symmetry has been an absolutely essential concept for physicists attempting to understand the structure of the physical world, this has not always been the case. More than forty years before the birth of supersymmetry, an English physicist by the name of Paul Dirac developed a theory that included what I consider to be the first quantum symmetry. Like supersymmetry, Dirac's symmetry is powerful and elegant, and even predicts the existence of new forms of matter. When Dirac's symmetry was confirmed experimentally, it represented the discovery of a lifetime. In the coming years, we may indeed witness another such momentous discovery in the form of supersymmetry.

A physical law must have
mathematical beauty.
—*Paul Dirac*

The Birth of a New Science

Rarely has humankind produced a specimen with the genius of Paul Adrien Maurice Dirac.

Once, when Paul Dirac was asked in what he believed, he simply declared that the laws of nature should be expressed in beautiful equations. A key figure in the history and development of twentieth-century physics, Dirac had a profound and deep respect for the elegance and simplicity built into the physical laws of our world. In a handful of fantastically insightful moments, he was able to see more deeply into the inner workings and structure of space and time than anyone had ever done before.

The 1920s were a period of chaos and revolution in physics. The classical laws had been considered firmly established since their introduction in the seventeenth century by Isaac Newton. Up until the twentieth century, there was no reason to think they would ever be overthrown. But in the first years and decades of the new century, it became increasingly clear that the smallest-known pieces of matter did not behave as predicted by the long-standing classical theory.

The way that electrons appeared to orbit around the nuclei of atoms seemed paradoxical and impossible to reconcile with the laws of physics as they were known. Furthermore, Albert Einstein had shown in 1905 that light—long understood to behave as waves—also acts as a collection of individual particles, or "quanta," as he called them. Whether light and matter consisted of waves, particles, or something else entirely different was not at all clear. Nothing made much sense. To understand these new and strange observations, the classical theory of physics—which had been successful for more than two centuries—would have to be replaced with something new. What was needed was a quantum theory of physics.

When Dirac first began studying mathematics and physics, there was no such theory of quantum physics. Bits and pieces of ideas had been put forth, some of which appeared promising and even seemed to solve some of the known paradoxes, but no complete theory had been discovered. Proposed solutions to one problem often introduced new difficulties in other areas, and nothing resembling a full picture of the phenomena was emerging.

Of Dirac's contributions to physics, his most important were his work on the quantum theory of matter and light. He would play a critical role in the development of the quantum theory. Well before he had begun to delve into the quantum nature of our world, however, Dirac had other scientific

Paul Dirac

interests. He had fallen in love with the beautiful ideas of Albert Einstein.

IT WAS WHILE STUDYING mathematics in the English city of Bristol that Dirac was first exposed to Einstein's theory of relativity. It is not hard to imagine why Dirac found this new theory so intriguing. Einstein's ideas had turned everything that had been thought about space, time, and gravity on its head. Einstein had single-handedly instigated a revolution.

In the second half of the nineteenth century, physicists studying electricity and magnetism had shown that light waves are, in fact, moving electric and magnetic fields—electromagnetic waves. Furthermore, according to electromagnetic theory, light waves must always travel through empty space at the same speed. The theory seemed to imply that everyone who ever measured the speed of a light wave would find the same speed—a universal value. This universality, however, leads to a number of paradoxical and perplexing conclusions.

Imagine, for example, that you are flying on a rocket, moving forward at a speed equal to half of the speed of light toward your friend, who is watching from the observation deck of a space station. As you approach your friend, you take out a flashlight and point it at her. Simple intuition would lead us to expect that your friend will see the light moving toward her at the speed of light (the speed of the beam of light from the flashlight) plus the speed of the rocket, for a total of one and a half times the speed of light. We simply add the velocities together: $1 + 0.5 = 1.5$. But light is *always* supposed to move at the same speed—the speed of light, never faster and never slower[1]—to all observers, including your

---

[1] This holds true only in empty space. The speed of light can be different in a medium, such as glass or water, for example.

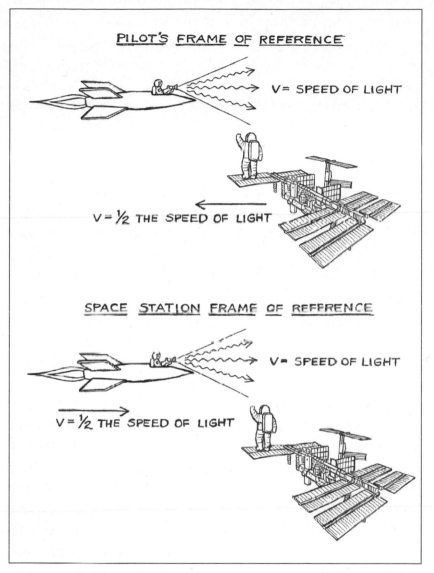

It would seem that light coming from an approaching object (such as a rocket) should appear to move more quickly than light from a stationary source. Light, however, is known always to travel at the same speed. This paradox is resolved by Einstein's theory of relativity.

stationary friend. Regardless of the speed of the rocket, she should see the light moving at the same speed that you do.

Something here is clearly wrong. This paradox led many physicists to think that the electromagnetic theory was somehow flawed, that light must actually be able to travel with a range of speeds, appearing differently to observers moving relative to each other. It seemed the most likely solution to almost everyone. But not to Einstein. In his 1905 paper, he proposed another solution—a very different solution.

When I said in the above example that we expected your friend, the stationary observer, to measure the beam of light to be moving toward her at one and a half times the speed that you observe light traveling at, I said we were led to this conclusion by "simple intuition." In this case, the principle our intuition is based upon is called the Galilean transformation—named after the great seventeenth-century astronomer and physicist.[2] According to the rules of Galilean transformations, velocities just add together—the speed of light from the flashlight plus the half of the speed of light from the rocket's motion equals one and a half times the speed of light, for example. This is exactly what our intuition leads us to believe should be true. The maxim of a good scientist, however, is always to be very suspicious of your intuition and what it leads you to believe, especially when its implications have not been experimentally confirmed. Common sense can, at times, be very deceptive.

In his theory of relativity, Einstein replaced the Galilean transformations with another set of rules called Lorentz transformations. For objects moving at speeds much slower than the speed of light, the Lorentz transformations act almost identically to the familiar

---

[2] It would probably be more accurate to say that the principle of the Galilean transformation is based upon our intuition, instead of the other way around. After all, human intuition was around well before Galileo.

Galilean rules. At much higher speeds, however, things behave very differently. Whereas we can write the Galilean rule for calculating the velocity your friend observes in the preceding example simply as $v_{rocket} + v_{light}$, the Lorentz transformations say something more subtle. Under the Lorentz tranformations, combining two speeds will never result in a speed greater than the speed of light, no matter how fast they are. For example, combining the speeds of two objects, each moving at half the speed of light, yields approximately 80 percent of the speed of light. The combination of 90 percent and 90 percent of the speed of light results in a velocity of just over 99 percent of the speed of light. Ninety-nine percent and 99 percent of the speed of light combine to 99.995 percent—still slightly below the speed of light.[3]

When we put into the Lorentz transformations half the speed of light for the velocity of the rocket, and the full speed of light for the velocity of the light from the flashlight, we find that the velocity of light as seen by your friend is not 150 percent of the speed of light, but the speed of light exactly. Einstein and his theory of relativity are able to restore the universality of the speed of light, as was required by the equations of electromagnetism. The same speed of light is observed by everyone, regardless of whether they are moving and what frame of reference they are in.

SO THE MATHEMATICS OF relativity works out in these examples, but what does it really mean? What causes our simple, Galilean intuition to break down at high velocities? What is going on here? What is behind it?

According to the theory of relativity, the speed of light is not only

---

[3] In case you are wondering, the speed of light has been measured to be about 299,792,458 meters (186,000 miles) per second, or 670, 616, 629 miles per hour.

the speed that light always travels at. It is also the fundamental and absolute speed limit built into the space and time of our universe.[4] Whenever you add energy into the motion of something, it speeds up. But no matter how much energy you put into something, it can never go faster than the speed of light—no way, no how, no matter what.

Consider once again the rules of the Lorentz tranformations. They predict that small velocities combine normally—by simple addition—but large velocities combine in such a way that they never exceed the speed of light. No matter how hard we try, we can't get a combined speed greater than the speed of light.

What is going on here is that space and time themselves are actually deformed to someone, or something, traveling at a speed close to the speed of light. Space is compressed—distances appear shorter—to someone who is moving quickly. Time, on the other hand, is stretched—lengths of time appear longer—to rapidly moving observers. Before Einstein, space and time were thought to be an unchanging and static background. After Einstein, space and time were understood to be dynamic and animate quantities that can grow and contract according to the motion of those who observe them.

One of the most interesting implications of Einstein's theory is that time travel is possible—at least time travel in the forward direction of time. Imagine that you get in a spaceship and fly at a speed just short of the speed of light—say, 99 percent of the speed of light—to the star closest to our own solar system, Alpha Centauri. At this speed, it would take you about four years to get there, and another four to get back. When you return, however, you would find

------

[4] In my office, I have a sign on the wall that my mom gave me for Christmas a few years ago. It reads "Speed Limit 670, 616, 629 MPH—It's not just a good idea, it's the law." Very clever, Mom.

that the calendars on Earth had not advanced eight years, but almost fifty-seven! The length of time that passes on Earth is stretched to the perspective of the moving observer—stretched from eight years to fifty-seven.

And just as time gets stretched, the lengths of moving objects get compressed. Imagine that as you fly your spaceship toward Alpha Centauri, you pass by your friend. Both of you measure the length of your ship. As far as you can tell, your ship is one hundred foot long—just as you measured it to be before you began your trip. To your stationary friend, however, your ship and everything in it is moving at 99 percent the speed of light and it appears much shorter—fourteen feet long instead of one hundred.

Although examples such as this can make Einstein's theory sound like wild and crazy science fiction, I assure you it is well tested and verified science. Over the past century, countless experiments have been performed to test the theory of relativity, none of which have ever found a departure from Einstein's predictions. Wild, yes, but also true.

After publishing his special theory of relativity in 1905 (along with four other revolutionary papers of the utmost importance), Einstein began working on a generalization of his theory—a theory of relativity that would also incorporate the force of gravity. In 1915, he finally succeeded, and published the first paper describing his general theory of relativity. When the general theory was experimentally confirmed for the first time in 1919, the whole world suddenly became very interested. The headlines of the London *Times* read, REVOLUTION IN SCIENCE—NEW THEORY OF THE UNIVERSE—NEWTONIAN IDEAS OVERTHROWN. The *New York Times* front page read, EINSTEIN THEORY TRIUMPHS, and quoted the great physicist J. J. Thomson as calling it "one of the greatest—perhaps the greatest—of achievements in the history of human thought." They also quoted Einstein himself saying that perhaps only twelve people in the world could comprehend his theory; he later denied ever saying any such thing.

As time went on and the general theory of relativity was further scrutinized, more and more bizarre predictions based upon it were made. Among the most perplexing and well-known examples is that of the existence of black holes—singular points in which space and time are warped to the most extreme limits. General relativity even enables us to understand the expansion of the universe itself, from an infinitely hot and dense Big Bang to the world we see around us today.

Einstein's theory has been described by many—including Paul Dirac—as the greatest scientific discovery ever made. Over the course of a single decade, Einstein's ideas had superseded the laws of motion *and* the theory of gravitation as they had been understood for the hundreds of years since Isaac Newton. The cornerstones of physics had been overthrown. But the revolution in physics was only beginning.

DIRAC POSSESSED A NUMBER of rather peculiar personality traits. He was a shy and sometimes meek individual, and he struggled in personal relationships throughout his life. Some have suggested that this was in large part due to the personality of Dirac's father, who himself was said to be unsociable, coarse, and cynically dark-natured. As a boy, Dirac ate dinner each night alone with his father, who allowed him only to speak French. Dirac's introverted tendencies, combined with his discomfort and unfamiliarity with the French language, led to many long and silent childhood evenings. Decades later, as a professor at Cambridge and later at the University of Florida, Dirac was known for his silence at seminars, rarely speaking or asking questions of the speaker. When he taught lecture courses on quantum physics, he would simply stand behind the podium and read—verbatim—from his own book on the sub-ject. If asked to explain or clarify a point in a lecture or seminar, Dirac would simply and patiently repeat the relevant portion of his speech, exactly as he had just said it. If still not understood, he would repeat it once again.[5]

As a boy, the introvert Dirac spent much of his time alone, out-doors in the English countryside. At this very young age, he devel-oped in his solitude a fascination with mathematical expressions, staring at them and exploring their meaning. He was attracted and deeply moved by the order and symmetry that he found in these equations.

I have on occasion wondered how different the history of science might have been if not for Dirac's peculiar nature. These quirks of his personality led to him developing an exceptional appreciation and tal-ent for mathematics. It was these abilities that enabled him to play a uniquely important role in the creation of what would eventually become a complete and self-consistent theory of quantum physics.

While Paul Dirac was studying mathematics in Bristol, Einstein's general theory was only a few years old. To the student Dirac, the lure of studying this new and exciting kind of physics was simply ir-resistible. Shortly after completing his mathematics course in 1923, he followed his heart and moved from Bristol to Cambridge. At Saint John's College at Cambridge was one of England's most prom-inent experts on Einstein's work—a mathematician by the name of Ebenezer Cunningham.

Ebenezer Cunningham had been interested in the theory of rela-tivity ever since Einstein's first paper on the topic was published in

---

[5] While at Cambridge, the university dons coined a new unit: the "dirac." The dirac was defined as the smallest measurable unit of conversation. A typical per-son might be described as a one-hundred-dirac speaker, while a more silent or shy person might be said to have only ten or twenty diracs.

1905. Over the years, he authored many papers on the subject of Einstein's theory. He also wrote *The Principle of Relativity,* the first book on relativity published in the English language. To the young Dirac, fascinated with Einstein's theory, Cunningham appeared the ideal choice of teacher and mentor.

When Dirac arrived at Cambridge, however, he learned that Professor Cunningham had very little time available, and was refusing to accept any new students. Forced to abandon the prospect of focusing his studies on relativity, at least for the time being, Dirac decided instead to pursue a course of study under the guidance of the mathematician Ralph Fowler. History would later call this seemingly unfortunate event very fortuitous indeed, as it was Fowler who introduced Dirac to the burgeoning field of quantum physics—the physics of the subatomic world. Very quickly—amazingly quickly, in fact—Dirac became a leading pioneer at the forefront of this new science. By the time he had received his PhD in 1926, he had become one of the world's most important physicists. He was only twenty-four years old at the time. A few years later, in 1933, he would be awarded the Nobel Prize in physics.

During Dirac's time as a student at Cambridge, the field of quantum physics consisted of only a handful of incomplete and often self-contradictory concepts and equations known to the small community of physicists and mathematicians working on the quantum theory. It was a mess—although an exciting mess. A number of important experiments had been carried out, pointing toward a very strange picture of how subatomic particles behave. Einstein had shown that light waves acted, at least in some ways, like particles. In contrast, Niels Bohr had shown that the behavior of electrons around the nuclei of hydrogen atoms is described better if the electrons are treated as waves rather than as particles. Despite these clues and insights, the observations had not yet been explained by any complete or consistent mathematical theory.

_____

BY THE MID-1920S, IT had become clear that new ways of thinking about physics would be needed if the world of atoms were ever to be understood. The more experiments that were performed to study these tiny pieces of matter, the more it became obvious that they do not behave at all in the way that the objects of our everyday experiences do. Atoms are not simply little versions of billiard balls, planets, or stones. They are very strange and counterintuitive beasts—unlike anything that had ever been imagined before.

By this time, experiments had determined that atoms consist of a small, positively charged center—a nucleus—surrounded by negatively charged and comparatively light electrons. A picture of a little solar system, with planetlike electrons orbiting around a sunlike nucleus, comes to mind. The problem with this picture, however, is that it simply doesn't work. Planets can stay in stable orbits around the Sun over billions and billions of years because they retain almost all of their energy as they travel. Electrically charged objects—such as electrons—radiate light whenever they move along curved trajectories. As they do this, they lose energy and, as a result, begin to move more slowly. An electron in an orbit around an atomic nucleus, therefore, would steadily lose its energy and quickly fall into the center of the atom. According to the laws of physics as they were understood in the first decades of the twentieth century, every electron in every atom in the universe should collapse into its nucleus in less than a billionth of a second. The universe and everything in it should have been entirely unstable.

Something—some process or mechanism—must be at work to prevent electrons from radiating and subsequently collapsing into the nuclei of atoms. Physicists were desperate for an idea that could explain the stability of atoms. In 1913, Niels Bohr proposed just such an idea.

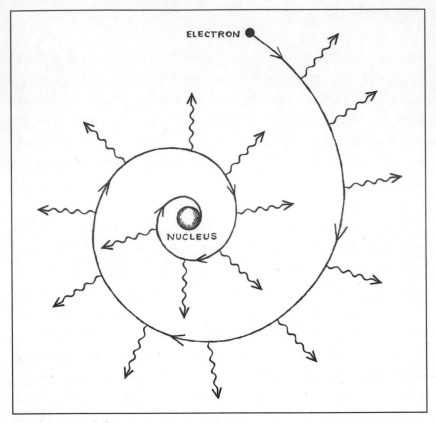

According to the laws of classical (prequantum) physics, electrons in orbits around the nuclei of atoms should continuously radiate light. As a result, they lose energy and quickly fall into their atom's nucleus.

Inspired by Einstein's conclusion that light waves behave in some ways like particles, Bohr entertained the possibility that particles—namely, electrons—might also act in some ways like waves. When you swish water around in your bathtub, you will find that some specific patterns of waves emerge. Certain sizes of waves form more easily than others, and last longer before dissipating. According to the hypothesis of Bohr, this set of wave patterns—called standing waves—were the only states in which electrons could ever be found around atoms. So, if we think about electrons as particles, they must steadily radiate light and lose energy; whereas, if we treat electrons as waves, then they can be found only in specific standing wave configurations, and cannot gradually radiate light and lose energy. Instead, they can only lose larger, discrete quantities of energy at a single time, causing them suddenly to "jump" from one standing wave state into another, lower-energy state.

Following his hypothesis, Bohr calculated the size of the various orbits in which electrons could be found around the simplest type of atom—the hydrogen atom. He found that the smallest orbit possible should be about a tenth of a billionth of a meter across. The second-smallest possible orbit is four times larger, followed by orbits nine, sixteen, and twenty-five times larger. Any electron in one of these larger orbits would almost instantly fall into the state requiring the least energy—the one with the smallest orbit—and stay there. No electron would, however, ever fall farther toward the center of the atom than this smallest orbit. Bohr's atom was indeed stable.

Whenever an electron jumps from one orbital state into a state of lower energy, an individual photon of light is emitted with a specific amount of energy. Bohr calculated the set of energies that these photons should be produced with in hydrogen atoms using his model, and compared it to the spectrum of light that had been observed from hydrogen. They matched surprisingly well. Bohr had not only stopped the atom from disintegrating, but had explained for the first time the spectrum of light produced by atoms.

If you are one of my more skeptical readers, by now you are asking, "But what the hell is this? An electron is a particle, not a wave!" Most people can understand what a water wave is, but an electron is a singular thing—something that should be at one place at one time, not a wave spread out over space and surrounding an atom. Even Bohr himself conceded that his model did not make much mechanical sense. Although the model did well in predicting and explaining some of the properties of atoms, it was not yet a real theory of how nature worked. Bohr merely had a glimpse of something that would turn out to be more wondrous and strange than he could possibly have imagined.

BETWEEN 1913, WHEN BOHR first described his idea of electron waves orbiting around atoms, and ten years later, when Paul Dirac began his studies in quantum physics, disappointingly little progress had been made in understanding the nature of atoms and light. This was all about to change. That change began in 1925, with the appearance of a paper written by the young German physicist Werner Heisenberg.

One day, late in the summer of 1925, Ralph Fowler was given a copy of an as of yet unpublished paper written by Heisenberg. After having a look himself, he decided to give the preprint to Dirac. Although Dirac did not immediately appreciate the importance of this new paper, after a short while he came to recognize that it contained within its pages the beginnings of an entirely new theory of quantum physics. Although Dirac was not the inventor of the Heisenberg approach to quantum mechanics, he was probably the first to understand fully what Heisenberg had done and what it meant; he probably appreciated some aspects of this work more than Heisenberg himself.

Like others, Heisenberg was very dissatisfied with the state of quantum physics as it stood in 1925. He considered the earlier work of Niels Bohr and others to be little more than speculation. Heisenberg's disregard for Bohr's model of the hydrogen atom was fueled in part by the fact that it predicted only *approximately* the spectrum of light that is emitted from hydrogen gas. Some of the observed frequencies were absent in Bohr's model, and others were shifted. Although Heisenberg thought that Bohr's ideas probably had something to do with how atoms work, he recognized that a real theory of quantum physics was still absent, and was desperately needed.

Heisenberg's paper, his first on the subject of quantum physics, was an attempt to construct new mathematical relationships describing the properties of quantum objects—electrons around an atom, for example. In classical—that is, prequantum—physics, these properties (the position or velocity of an object, for example) could each be simply described by a number. I work at the Fermi National Accelerator Laboratory, which is forty miles west of Chicago. And I drive at a speed of sixty-five miles per hour to get there each morning. Heisenberg's paper, however, described these properties in a new mathematical language. Instead of simply using individual numbers to denote properties such as position and velocity, Heisenberg used more complex mathematical structures known to mathematicians as matrices.

Dirac realized almost immediately that these mathematical matrices of Heisenberg possessed very special and interesting properties. Namely, they did not commute. Ordinary numbers do commute, meaning that $A \times B$ is identical to $B \times A$ for any two numbers, A and B. We normally just take this rule for granted; so much so that it is probably hard for you to imagine something for which $A \times B$ is not equal to $B \times A$.[6]

As Dirac was quick to realize, it was this characteristic of Heisenberg's mathematics from which its most interesting and important results come. Heisenberg, in contrast, had tried hard to downplay this feature of his paper. He mentioned the property only once, in a single sentence, buried deep within the text. To illustrate his method,

---

[6] Although unknown to Heisenberg, Dirac, and the overwhelming majority of physicists in the 1920s, matrices had been studied by mathematicians for some time. Today, matrices are widely used in physics, and are taught as a standard subject to virtually every university student majoring in mathematics or physics.

he carefully selected a special case example that *did* obey $A \times B = B \times A$, despite the fact that this was not generally expected to be true in Heisenberg's formalism. In other words, Heisenberg was afraid that this strange property would spoil his otherwise beautiful system. Dirac, on the other hand, was not in the least bit afraid to follow the mathematics to wherever it would lead him. Dirac felt a profound confidence in, and even a loyalty toward, mathematics. Many years after the quantum revolution, near the end of his life, Dirac expressed it this way:

"One should allow oneself to be led in the direction which the mathematics suggests . . . one even must follow up a mathematical idea and see what its consequences are, even though one gets led to a domain which is completely foreign to what one started with. Mathematics can lead us in a direction we would not take if we only followed up physical ideas by themselves."[7]

To Dirac, mathematics was a tool capable of revelation—something that should be trusted and pursued devotedly. Whereas Heisenberg was uncertain and wary of these strange properties that had emerged from the mathematics, Dirac delved into them seriously and without apprehension. It was this unhindered attitude that allowed Dirac to recognize that it was precisely these perplexing mathematical properties that made Heisenberg's work all the more beautiful.

PAUL DIRAC WAS NOT the only physicist to have insights into the underpinnings of Heisenberg's quantum relationships. While Dirac was investigating the mathematics behind Heisenberg's paper in Cambridge, Max Born—Heisenberg's doctoral adviser—was doing the same in Göttingen, Germany. Over the next several months, a

---

[7] From *The Second Creation,* by Robert P. Crease and Charles C. Mann (Macmillan, 1986).

series of papers by Dirac, Born, Heisenberg, and others were written, further exploring the properties and structure of the new quantum theory. This small group of physicists also began corresponding with each other. Dirac even traveled to Copenhagen and later Göttingen, where he met and exchanged ideas with Heisenberg, Bohr, Born, and others. Over time, a much deeper understanding of the quantum world began to emerge.

Only months after Dirac and Born had started working on Heisenberg's quantum theory, a new series of papers began to appear by the German physicist Erwin Schrödinger. These papers introduced yet another approach to calculating the behavior of quantum systems, such as the hydrogen atom. The central equation in the papers—known today as the Schrödinger equation—describes electrons and other particles as waves, reminiscent of the work of Bohr. To some physicists, this new theory seemed destined to replace the work of Heisenberg's almost before it had seen the light of day. Heisenberg and his supporters, however, were not going to give up without a fight.

One advantage possessed by Schrödinger's new approach was that it was comprehensible to most physicists—far more so than Heisenberg's matrices. After all, the physicists of the 1920s had all studied the behavior of waves in school and were generally familiar with the mathematics needed to describe them. Almost no physicists had studied the properties of matrices. This put Heisenberg almost immediately on the defensive in the debate between the two quantum theories. This debate quickly became very heated, with the stakes being seen as a likely Nobel Prize and a prominent place in the history of science. The supporters of Schrödinger's wave mechanics and Heisenberg's matrix mechanics each seemed absolutely confident that their theory was the correct one. As it turns out, they were both right.

The debate came to a crashing halt with another brilliant insight from Dirac. Although it had not been at all clear before, Dirac was able to demonstrate that the two seemingly very different systems of Schrödinger and Heisenberg were, beneath it all, actually the same theory. Dirac even went so far as to show how one could transform the language of each system into the other. Once the dust settled, it was accepted that the mathematics of the two quantum theories actually describes the same phenomena, and make the same predictions. Today, they are taught side by side in university courses as alternative, but equally valid, descriptions of the quantum world. They are a single theory of quantum mechanics.

The development of the new quantum theory as constructed by Heisenberg, Schrödinger, Dirac, and others marked the emergence of a new generation of quantum physicists. The old guard—led by Einstein and Bohr—had been superseded by this young and inspired group. In 1926, Dirac and Heisenberg were twenty-four and twenty-five years old, respectively, Schrödinger being the elder of the generation at the age of thirty-nine.[8]

Matrices, the mathematical structures appearing in Heisenberg's paper, can be thought of as something similar to lists—lists of the likelihood of a particle being observed in a specific place, or with a specific velocity, for example. In the case of the hydrogen atom, Heisenberg's matrices list the probabilities of an electron being found in each of the possible orbits, similar to those found by Niels Bohr in his wave model. Although Bohr's description of atomic

---

[8] I've always found it strange that so large a proportion of photographs and portrayals of great physicists are of elderly men, especially as presented in popular culture. In reality, a very large fraction of the most important scientific advances have been made by much younger individuals. Next time you see a picture of an elderly Einstein with his billowing gray hair, remember that he did his most important work on quantum physics and invented the special theory of relativity by the time he was twenty-six. His single greatest achievement, the general theory of relativity, was published when he was thirty-six years old.

orbits did not turn out to be entirely correct, many features of his model survived intact within both Heisenberg's and Schrödinger's versions of the quantum theory.

When the variables denoting position and velocity are replaced by matrices in the classical—prequantum—equations of motion, the true underlying strangeness of quantum physics begins to emerge. As I said before, much of this strangeness comes directly from the inability of Heisenberg's matrices to commute (A × B being not equal to B × A). Because of this property, it can be shown that no quantum state can exist that has exactly one unique value for position *and* one unique value for velocity. An object in precisely one place with precisely one speed cannot exist. The more specific or precisely determined the location of an object is, the less precisely its velocity can be defined, and vice versa. This fuzziness is encapsulated in what is known as the Heisenberg uncertainty principle. The quantum world is not the place we intuitively understand from our everyday experiences. The quantum world is blurry and bizarre.

Imagine an electron traveling around an atom. Our intuition leads us to think that if we had precise enough tools (rulers, calipers, microscopes) we could measure the location and the velocity of the electron to whatever precision we wanted to. This is not the case. There are limits to how precisely defined these quantities are. If the electron's location is known to be somewhere inside the smallest hydrogen atom orbit as calculated by Bohr—about a tenth of a billionth of a meter across—then the Heisenberg uncertainty principle says that its velocity cannot possibly be known to better accuracy than within a few hundred thousand meters per second. Considering that the average velocity you can expect to find for an electron in a hydrogen atom is not much larger than this, it means that the velocity of the electron is almost entirely undetermined. It could be found to be moving at many millions of meters per second, or not moving at all, or moving with some other velocity in between.

This is not the result of a limitation of our measurement tools or their accuracy. It is not even the result of a limitation of our knowledge. Even if you somehow learned everything there is to know about an electron, you would not and can never know precisely its location and velocity. Even an omnipotent, all-knowing god would not and could not know. An electron is simply not in a single place and does not move at a single speed. It fills the space around the center of the atom, being spread out over that volume of space. Similarly, it simultaneously takes on a range of velocities.

Although we do not notice it, the same strange properties are also present in our macroscopic world. If you know the location of a baseball to within a millimeter, its velocity cannot possibly be determined with a better accuracy than within about a billionth of a billionth of a trillionth of a meter per second—an absolutely minuscule degree of uncertainty. If you are driving down the road and know your car's speed to within one mile per hour, its location is smeared out over about a trillionth of a trillionth of a trillionth of a meter. Once again, quantum uncertainty is irrelevantly small to the objects encountered in our daily lives.

But as one considers smaller and smaller objects, the strangeness of quantum physics becomes more and more apparent. The behavior of individual particles—electrons, photons, and so on—is somewhere between very hard and impossible for us to envision. No matter how much we try, we can never picture clearly how an electron behaves or what it looks like. Electrons act nothing like the baseballs, automobiles, and other objects we are familiar with. Our bodies—including our brains, eyes, and ears—are incredibly well suited for perceiving and understanding the world around us. But they have absolutely no reason to be capable of understanding the quantum world. Millions of years of natural selection and evolution have made us capable of instantly and instinctively identifying colors, shapes, movement, and other patterns in what we see and experience. These tools—so very well adapted to our common experiences—do not enable us to identify with the quantum world. The best we can do is to experience the quantum universe through analogy to the experiences of our world. As the great physicist Richard Feynman once said, "It is safe to say that nobody understands quantum mechanics." In a sense, he is right.

THE WORK OF SCHRÖDINGER, Heisenberg, Dirac, and Born led to the birth of a real theory of quantum physics. They were certainly proud parents, and with good reason. They had before them the beginnings of a beautiful and wondrous theory of nature. But as any parent knows, a child has a long way to go before it reaches maturity. Quantum physics had not yet grown up.

> **A great deal of my work is just playing with equations and seeing what they give. I don't suppose that applies so much to other physicists; I think it's a peculiarity of myself that I like to play about with equations, just looking for beautiful mathematical relations which maybe don't have any physical meaning at all. Sometimes they do.**
> —*Paul Dirac*[1]

Dirac's Symmetry

The history of science is a long and varied story. Most of this story consists of tales of clever individuals who saw beyond the existing understanding of our world. These scientists noticed mysteries that had not yet been explained and followed them to the discovery of a new principle or physical law. Galileo demonstrated that gravity pulls all objects toward Earth at the same rate, whereas his predecessors followed Aristotle in their belief that heavier objects must fall more rapidly. Isaac Newton determined that planetary orbits could be explained if gravity pulled all objects together with a force that becomes weaker with the square of their distance from each other. Niels Bohr realized that if you treated electrons like waves, the spectrum of light observed from hydrogen could be better understood and predicted. These discoveries each represented great leaps forward in the way we understood our world.

---

[1] From *The Second Creation,* by Robert P. Crease and Charles C. Mann (Macmillan, 1986).

There is, however, an even more fantastic and exceptional kind of scientific advance that sometimes occurs. On very special occasions, pure reason and thought leads us to understand things about our world even before they are observed—the opposite to the usual order in which most scientific advances are made. In these instances, it becomes apparent not only how our world is, but also how it has to be. In these awesome and overwhelming moments, one suddenly appreciates what the mathematics behind our world demands of itself.

These insights into the underlying logic and very structure of our universe are to some physicists much like moments of divine revelation to the theologically minded. As far as I am aware, divine revelation is described in the tales, myths, and traditions of all world religions. Whether it is God speaking to Moses at the top of Mount Sinai, or the mathematics of our world becoming understood for the first time, these are special moments—moments in which a newly revealed truth comes forth from the foundation of the universe itself. Although few scientists would describe these instances of mathematical insight as divine, there is no denying that they are moments of revelation.

One of the most remarkable moments of mathematical revelation ever to take place was experienced by Paul Dirac along his journey toward developing the quantum theory. His insights into the quantum nature of our world would reveal much that had never been seen before, nor even imagined. Matter, he would learn, must exist in other forms than were then known. Dirac's mathematics would force him to conclude that matter can be both created and destroyed, and could spontaneously come into and fall out of existence.

What Dirac discovered is a symmetry that is built into the very fabric of our world—a relationship between what we call matter and something else, the opposite of matter, called antimatter. This is the first of our world's quantum symmetries to become understood and appreciated. Today, we know of many other examples of such symmetries. These symmetries demand things of our world. Supersymmetry demands that if matter is to exist, then force must exist alongside it. According to Dirac's symmetry, nature demands that if matter exists, then antimatter must exist as well. And exist it does. It simply could not be any other way.

THE DISCOVERY OF QUANTUM theory was an accomplishment of the absolutely highest order. It revealed that matter and light behave in ways that could not have been imagined by the physicists of previous generations. The world is a strange and amazing place. The quantum theory is a magnificent testament to this fact.

The new quantum theory as formulated by Heisenberg, Schrödinger, Dirac, and others was, however, not complete. A close inspection of the equations revealed the limitations of its applicability and ultimately its accuracy. In particular, the quantum theory was not designed in a way that could be consistent with Einstein's special theory of relativity.[2] The quantum theory as it emerged in the mid-1920s was only an approximation, and only revealed a glimpse of how rich and profound the quantum world truly is. Dirac took it upon himself to discover how Einstein's relativity could be incorporated into the quantum theory of matter and light. Dirac set out to discover the complete quantum theory—a relativistic quantum theory.

According to Einstein's relativity, energy comes in two forms. The first of these forms is the energy associated with motion, called kinetic energy. A baseball thrown at a hundred miles per hour has more energy than a baseball sitting still. The second form of energy is what we

---

[2] This is not to be confused with the general theory of relativity. Even today, no one has found a way of combining general relativity with quantum physics.

call mass. This notion is encapsulated in Einstein's most famous equation, $E=mc^2$. This equation simply says that mass is one form of energy. In fact, mass can contain a lot of energy. The mass of a baseball constitutes about 100 trillion times more energy than the kinetic energy of a major-league fastball. In order for a baseball to have as much energy in the form of motion as it does in its mass, it would have to be traveling at almost 90 percent of the speed of light.

The Schrödinger equation was built upon the old, pre-Einsteinian description of energy, and doesn't account for mass in the way that the theory of relativity demands. The Schrödinger equation describes the energy of a particle's motion but ignores the energy associated with its mass. Dirac, Schrödinger, and others were not surprised that this treatment works well when applied to particles moving somewhat slowly—far below the speed of light. But they also knew that the quantum theory, as it was formulated, would break down as particles became more energetic and moved more rapidly.

Schrödinger constructed his quantum equation by replacing the variables for position and momentum in the classical equation of motion with mathematical entities called operators—in much the same way that Heisenberg replaced these variables with matrices. Dirac began his quest for a more complete quantum theory by trying to do the same thing. But instead of starting where Schrödinger had, with the old Newtonian equation, Dirac started with Einstein's relativistic equation of energy. It was the first step toward understanding the quantum nature of matter in a way that no one had ever done before.

DIRAC WAS NOT THE first to embark upon a quest to formulate a relativistic quantum theory. Erwin Schrödinger, even before he had published his famous equation, had also attempted to make quantum physics consistent with Einstein's theory. Just as he did with the Newtonian equation, Schrödinger took Einstein's basic equation of energy and replaced quantities such as momentum and energy with mathematical operators. But when he did this, the results seemed nonsensical.

Schrödinger's original equation could be used to calculate the probabilities of finding a particle in various places, or with various velocities. The version of his equation that was built upon Einstein's theory, however, produced very strange probabilities. In particular, it was possible that the probabilities of finding a given particle in the sum of all possible places could, in some cases, add up to more than 100 percent. To Schrödinger, it was as if he had calculated that the probability of flipping a coin and getting heads was 75 percent and at the same time the probability of getting tails was 75 percent. It just seemed to make no sense.

After giving up on building a quantum equation that was consistent with Einstein's, Schrödinger settled for an approximate equation that worked as long as the particles were moving slowly compared to the speed of light; this is the famous equation that bears his name. The relativistic quantum equation that Schrödinger abandoned was later rediscovered and published by Oskar Klein and Walter Gordon, but still without a solution to the issues with probabilities.[3] A couple of years later, when Dirac began to think about the problems involved with building a relativistic quantum theory, he picked up Schröding-er's quest where it had been abandoned. Dirac wandered blindly into the mathematical wilderness that lay ahead.

Dirac, as I've said before, possessed a rare genius for mathematics, with exceptional intuition and creativity. Beginning with the same tools as Schrödinger—Einstein's energy equation and the basic quantum operators—Dirac found much, much more. In addition

---

[3] Today, this is known as the Klein-Gordon equation.

to the equation that Schrödinger and later Klein and Gordon derived, Dirac found another equation—one that did not lead to the same problems.

This new equation—known as the Dirac equation—was considerably more complicated than Schrödinger's formula. Whereas every undergraduate physics student learns to use Schrödinger's equation—the one that works for slowly moving particles—the Dirac equation is usually not studied by physics students until graduate school, and even then in most cases only by students specializing their studies in particle or nuclear physics. The mathematics involved in solving Dirac's equation can be a frustrating mess. At first glance, Dirac's equation appears as anything but clear and simple. Beneath it all, however, its mathematical beauty is undeniable.

When Dirac found the basic solutions to his new equation, a strange property emerged. As he expected, he found that particles—such as electrons—could exist in a wide range of different states. They could have small or large quantities of energy, and be moving slowly or quickly. The strange thing that he found from his equation was that for every possible state with a given energy, there was also a possible state with the opposite amount of energy—a *negative* amount of energy.

This result appeared to make no sense at all. A particle cannot have a negative energy, no more than you can drive your car at a speed of negative sixty miles per hour; there is a good reason why your speedometer stops at zero. Physicists sometimes talk about negative energies when comparing something to a more energetic object, but that is just an energy that is negative relative to another. Dirac's equation seemed to predict particles with a negative absolute energy.

If a less-imaginative physicist had reached this point in his pursuit of the quantum theory, he probably would have quit, as Schrödinger had done. Dirac did no such thing. Instead, he began to try to resolve or remove the weird negative-energy solutions of his equation.

He was not about to abandon the path that the mathematics was revealing to him—not without a fight, at least.

The most obvious approach to try was simply to forbid the negative states from existing. By doing this, one could use just the equation's positive-energy solutions to make predictions about how particles behave in our world. This, however, didn't work. The complete set of all possible states—both positive and negative—need to be present in order for the theory to be self-consistent. The unwanted negative-energy states would have to be dealt with somehow, and couldn't just be dismissed. This forced Dirac to contemplate a far more radical idea to solve the problem of negative-energy states. Out of desperation, Dirac turned to something known as "hole theory."

Imagine for a moment that all of Dirac's states—with both positive and negative energies—are real and do exist as possible configurations that an electron could be found in. Any electron starting in a positive-energy state would quickly fall into increasingly lower-energy states, like boulders rolling down a mountain, or electrons in Bohr's atom moving into orbits as close to the nucleus as possible. Nature always acts to minimize the energy possessed by an object. Because there is no minimum energy state according to Dirac's equation—the solutions go down forever, to negative infinity—an electron would never stop losing energy. But this isn't at all like what happens in the world we experience; rather, we see electrons with positive energy all around us. This infinite descent simply cannot be allowed to take place. But what could stop it?

In a bold and brazen move, Dirac hypothesized that electrons do not fall into the oblivion of negative-energy states because all of the negative-energy states are already full. A few years earlier, in 1925, the physicist Wolfgang Pauli had shown that no two electrons could occupy the same quantum state at the same time—a rule known as the exclusion principle. Dirac postulated that there must exist all around us a vacuum of unseen electrons filling every single one of the negative-energy states predicted by his equation. This infinite "sea" of negative-energy electrons fills all negative-energy states, and thus—according to Pauli's exclusion principle—stops any positive-energy electron from falling into them.

According to Dirac's picture, the vacuum—what we call empty space—is not really empty at all. Instead, it is the arrangement in which all of the possible negative-energy states, and none of the positive-energy states, are full. But there are consequences of this interpretation. For example, ordinary particles should be able to interact with the negative-energy particles making up the vacuum of empty space. Imagine, for example, that a photon of light comes along and strikes one of the negative-energy electrons in the vacuum, in the process transferring enough energy to it to give it a net positive energy. This leaves us with a positive-energy electron *and* an absence of a negative-energy electron where one used to be in the normal vacuum—a "hole" in the vacuum.

This hole that appears in the vacuum merits a closer look. Compared to the normal, entirely full state of the vacuum, the vacuum with a hole in it is short one electron and therefore has less negative electric charge. So instead of thinking of a hole as a shortage of negative charge, one can think of it as a presence of positive charge. In other words, a photon knocking a negative-energy electron from the vacuum sea into a positive-energy state can be thought of as the photon creating two particles, one negatively charged and one positively charged, but each with positive energy. According to this hole interpretation of the solution to Dirac's equation, the nonsensical negative-energy electron states are not negative energy at all. Instead, they are positive-energy states, with positive electric charge.

DIRAC'S HOLE THEORY INTERPRETATION was not without its problems. To begin with, the infinite sea of negatively charged

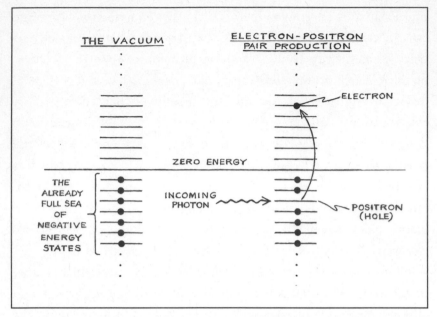

In Dirac's hole theory, a photon can interact with an electron in the vacuum sea, transferring energy to it in the process. The empty quantum state that is left behind–the "hole"–appears to us to be a particle similar to an electron, but with opposite electric charge and other properties. These holes in his theory led Dirac to predict the existence of antimatter.

particles that makes up the vacuum—the absence of any holes—is not felt. After all, empty space doesn't seem to have electric charge. To avoid this problem, there must be something about the underlying vacuum that makes this go away—perhaps the "bare vacuum" beneath it all has a positive charge that cancels out Dirac's sea, for example. This was not a particularly elegant solution to the problem, but at least it was a way to sweep it under the rug for the time being.

On top of this problem, Dirac's equation seemed to require that the positively charged states—the holes—should be particles with the same mass as electrons. At the time of Dirac's calculations, however, no positively charged particles had ever been observed besides protons.[4] Protons, however, are about two thousand times heavier than electrons. They didn't seem to fit the role of Dirac's quantum holes. Nonetheless, Dirac speculated that protons could somehow be the holes in his sea. This, in turn, led to entirely new problems. Just as a pair of positive and negative particles could be brought into existence together by moving a particle from a negative-energy state to a positive one, this process should be able to happen in reverse, causing two such particles to destroy each other. If protons were in fact the holes in Dirac's theory, then the electrons and protons bound up together in atoms would instantly annihilate each other. In Dirac's theory, atoms seemed to be unstable. Dirac's theory appeared very troubled.

Although Dirac did not realize it immediately, his equation itself contained the solution to its problems. If one takes a somewhat open-minded look at Dirac's equation, one conclusion becomes apparent: there really must exist a type of particle which is exactly like an electron, but with a positive rather than negative electric charge. Dirac was reluctant to come out and say it directly, but his equation had predicted the existence of a new kind of particle—in fact, a new kind of matter.

AT ABOUT THE TIME that Dirac was inventing his equation, another new subfield of science was being born. Radiation—energetic particles—had been observed to be striking Earth from outer space. Although the first evidence of this otherworldly radiation was seen in 1912, it was years before much was known about it. Beginning in the late 1920s, a wide range of experiments were designed

---

[4] There were also many positively charged (ionized) atoms and molecules known to exist at the time, but each of these has a net electric charge only because they contain more protons than electrons.

and carried out to study these mysterious "cosmic rays." One of the physicists who set out to study the cosmic radiation was Carl Anderson.

Anderson was born in New York City, but later moved to Pasadena to study physics and engineering at Caltech under the tutelage of the famous physicist Robert Millikan. Anderson stayed at Caltech through his entire life, doing both his undergraduate and graduate study there, and afterward becoming a member of the faculty. In 1932, only two years after earning his doctorate, Anderson made the first of the great discoveries of his career. It was this discovery that would earn him the 1936 Nobel Prize in physics.

One of the tools Anderson and others used for studying cosmic rays was a device known as a cloud chamber. A cloud chamber is essentially a container enclosing a volume of gas, kept near its condensation point. When an electrically charged particle passes through the gas in the chamber, droplets form along the particle's path, leaving a trail behind it. With a magnetic field, the charged particles can be made to move along curved trajectories. By studying the shape of a track left in the chamber, it is possible to determine the mass and the electric charge of the particle passing through.

One day, Anderson saw a different kind of track in his cloud chamber than he was accustomed to. It looked a lot like the track an electron would make, but it curved under the influence of the magnetic field in the opposite direction. What his cloud chamber had recorded was something with the same mass as an electron, but with opposite electric charge—a positively charged electron, or positron.

Dirac must have been ecstatic. The existence of positrons was exactly what was needed to put his equation and hole theory back in order. With the discovery of this new kind of matter, many of the troubling contradictions of Dirac's relativistic quantum theory—such as the instability of atoms—simply evaporated.

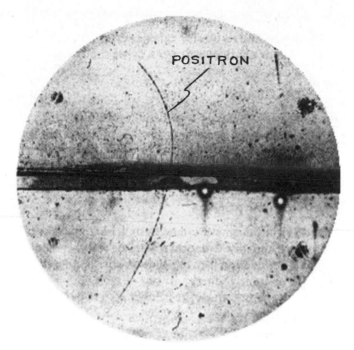

A photograph of a positron appearing in Carl Anderson's cloud chamber. By studying the trajectory of a particle through the chamber's magnetic field, its mass and electric charge can be determined. For the particle in the photograph above, its trajectory has the same degree of curvature as would be expected for an electron, but turning in the opposite direction. The particle, therefore, has the same mass as an electron, but with a positive charge—a positron. *Reprinted with permission from Carl D. Anderson,* Physical Review 43 (1933): 491. *Copyright © 2004 by the American Physical Society.*

In his original paper on his relativistic quantum theory, Dirac never explicitly predicted that positrons should exist, being afraid that it would be seen as nothing but wild speculation. Two years after his first paper on the subject, however, he finally acknowledged that his equation clearly calls for positrons. To Dirac, there was no way around this requirement of his equation. Only a year after Dirac made this prediction, Anderson discovered the positron. If Dirac had waited even a year longer to admit to his theory's consequences, history may have viewed his greatest achievement in a very different light. When asked why he had not predicted the positron's existence earlier, Dirac simply replied, "Pure cowardice."

IN THE YEARS FOLLOWING the confirmation of the Dirac equation, a series of other developments took place in the construction and refinement of the quantum theory. Among other advances, it became understood that the Dirac equation is applicable not only to electrons, but to all fermion particles. Other kinds of particles follow other sets of rules, however. Bosons obey the rules of other equations, one of which is the Klein-Gordon equation—the one that Erwin Schrödinger had previously discarded. By the mid-1930s, the problems that had once led to the abandonment of this equation had been resolved. In essence, the solution resulted from the new understanding that particles could be created and destroyed. If it was no longer assumed that the number of particles had to remain the same in a given process, then the problems with probabilities found in this equation simply disappeared.

It took some time, but gradually Dirac's relativistic quantum theory grew into something called quantum field theory. By the mid-1930s, quantum field theory was well on its way, but the later refinements made by the postwar generation of physicists—including Richard Feynman, Julian Schwinger, and Freeman Dyson—completed it and gave it its modern form. In quantum field theory, particles alone are not the whole story of matter's existence. In this more modern and complete picture, Dirac's holes are only a helpful way of conceptualizing the behavior of our universe—an illustrative analogy, rather than a literal description. Instead of particles and holes, we now understand Dirac's equation to describe quantum fields extending over space, frothing and bubbling, bringing forth and destroying particles.

Particles are the manifestations of these underlying quantum fields. When something feels the influence of an electric or magnetic field, it is in fact experiencing photons traveling between charged particles and communicating the electromagnetic force. If photons did not exist, then neither would the effects of electricity or magnetism. The electromagnetic force is no more and no less than photons being passed between particles through space.

Experimental tests of quantum field theory have been mind-bogglingly successful. Every single test that has ever been conducted has confirmed its accuracy. The most precisely measured quantities agree with the theoretical predictions of quantum field theory to a staggering degree. The magnetic moment of the electron, for example, has been measured to within about one part in a trillion—and is in perfect agreement with the predicted value.[5] In the many decades since its introduction, there has never been an experimental result that has demonstrated a failure of quantum field theory.

BUILT INTO THE VERY foundation of quantum field theory is the symmetry introduced by Dirac. Just as the Dirac equation requires that positrons must exist along with electrons, the same symmetry can be applied to all varieties of particles. In order for electrons, protons, neutrons, quarks, neutrinos, and so on to exist, their antimatter counterparts must also exist—positrons, antiprotons, antineutrons, antiquarks, antineutrinos, and so on. Each of these antimatter particles has precisely the same mass as their matter counterparts, but opposite values for their electric charge and other properties.

Even boson particles—which do not obey the Dirac equation—are required to obey the matter-antimatter symmetry. The boson called the W comes in both positive and negative versions, for example.

---

[5] If you don't know what a magnetic moment is, don't worry. All that is important to understand here is that the experimental tests of quantum field theory have been extremely successful.

Even photons, which have no electric charge or other properties to make opposite, obey this symmetry. The photon's antimatter counterpart is simply itself.[6] Nowhere in nature has there ever been observed a type of particle without a corresponding antiparticle.

Despite the existence of antimatter, you might have noticed that there isn't much of it in our world. The room you are in, the chair you are sitting on, the air you are breathing, and the planet we live on are all made almost entirely of matter and almost no antimatter. There is a good reason for this. Whenever a particle of antimatter approaches its matter particle counterpart, the two particles are almost instantly annihilated—leaving behind only energy in the form of photons of light. In fact, according to Einstein's equation $E=mc^2$, a lot of energy is released in this kind of process. A single gram of antimatter, if brought into contact with ordinary matter, would release roughly as much energy as the Hiroshima and Nagasaki bombs put together.

This idea is the premise of Dan Brown's best-selling novel *Angels & Demons*. In this story, the antagonist gets his hands on a canister of antimatter, which he intends to use to destroy a city by allowing it to come into contact with ordinary matter.[7] Fortunately, the scenario of an antimatter-wielding terrorist is fairly far-fetched. In order for antimatter to be used as a weapon of mass destruction, it would have to be created and then contained without any contact whatsoever with matter particles. You might have noticed, however, that there are matter particles almost everywhere, and antimatter particles almost nowhere. All of the antimatter physicists have ever produced in experiments would barely provide enough energy to power a lightbulb for a few minutes—hardly the kind of threat envisioned by

---

[6] This is because the photon's electric charge and other quantum properties are each exactly zero. Since the negative value of the number zero is simply zero, if you make all of the quantum properties of a photon opposite, you still have a photon.
[7] I won't ruin the story by telling you who the antagonist turns out to be.

Dan Brown. Furthermore, even if somehow you did manage to create a dangerous quantity of antimatter, in order for it not to immediately explode, it would have to be kept insulated and out of contact with the surrounding world—perhaps using a carefully controlled magnetic field. It would be extremely difficult for a dangerous quantity of antimatter to be kept isolated in this way.

There are also far more practical—and benevolent—ways in which antimatter can be used. The most striking application is found in positron-emission tomography (PET) scans. This technique is used to take pictures of your brain by first injecting into your body radioactive materials that release positrons. The positrons quickly come into contact with ordinary electrons in your brain. When that happens, the particles destroy each other and, in the process, give off energetic and distinctive photons of light. It is these photons that are actually detected by the scan.

In the somewhat distant future, if substantial amounts of antimatter could be safely stored, it could also be used as a very light and portable energy source. Future spacecraft may even use antimatter batteries to propel themselves, for example. These batteries would very gradually release particles of antimatter, which would instantly interact with matter and release energy, thus accelerating the craft. Gram for gram, no fuel contains as much usable energy as antimatter.

IN DIRAC'S DAY, THE concepts behind quantum physics were not often thought of in terms of symmetry. I don't know if even Dirac thought about the relationship between matter and antimatter in terms of symmetry. Since Dirac's day, however, much has changed.

Over the decades following the birth of quantum field theory, quantum physics has gradually transformed and grown into the field of particle physics. Symmetry is at the very heart of particle physics.

The varieties of particles that can exist, the nature of their interactions with each other and their other properties are all understood in terms of the symmetries they obey. In 1932, the only types of particles known to exist were electrons, positrons, protons, neutrons, and photons. Since then, dozens and dozens of new forms of matter have been discovered. Despite this proliferation of particle species, much of particle physics can be understood in terms of a few basic symmetries. These symmetries, and the particles that they describe, are known today as the Standard Model. Although the path from Dirac's work to the Standard Model is a long and meandering one, it has ultimately led to the single most successful theory that science has ever produced.

> One thing that makes the adventure of working
> in our field particularly rewarding, especially
> in attempting to improve the theory, is that . . .
> a chief criterion for the selection of a correct
> hypothesis . . . seems to be the criterion of
> beauty, simplicity, or elegance.
> —*Murray Gell-Mann*

From Simplicity to Chaos . . .
and Back Again

With the existence of the positron verified, five forms of matter had become known: protons, neutrons, electrons, positrons, and photons. This short list describes a great deal of what is seen in our world. For one thing, every last one of the chemical elements of the periodic table is a combination of protons, neutrons, and electrons bound up together. To catalog all of the atoms and molecules observed in the world once required a vast list that just kept getting longer and longer. Prior to the eighteenth century, about a dozen varieties of chemical elements were known, most of which were discovered in ancient times. By the beginning of the twentieth century, however, more than eighty elements—and many more molecules—had been discovered. The long list of atoms and molecules made chemistry seem to many like a rather complicated and inelegant science—chemical taxonomy. But hidden beneath that very long list, all of chemistry ultimately comes down to nothing more than protons, neutrons, and electrons. That realization transformed our understanding of matter. From a countless number of atoms and molecules to three simple underlying components, humankind had taken a great leap forward in understanding our world.

But this simple picture of our world did not remain in place for long. Physicists who looked carefully at the state of quantum physics in the mid-1930s knew that there were holes—gaping holes—in their understanding of matter. All signs pointed toward the necessary existence of other varieties of matter and force. The simple description of our world as only protons, neutrons, electrons, positrons, and photons was soon to be overthrown.

Dirac's theory itself required that all varieties of fermions must have antimatter counterparts. But in the 1930s, only the positron had been discovered. If Dirac's theory were truly correct, there must also be antiprotons and antineutrons in the world. But uncovering the existence of the other forms of antimatter was only the beginning. In addition to these particles, the still-developing quantum field theory seemed to insist that even more forms of matter must exist as well—particles intimately tied to the forces of our world. In quantum field theory, the electromagnetic force is the manifestation of photons being passed back and forth through space, transferring energy and momentum—pushing and pulling—between electrically charged particles. Well, electromagnetism is not the only force in our world. And in a world with other types of force, there must also be other types of particles to communicate those forces—indeed, to bring those forces into being.

Today, we know of four different forces that are present in our universe—gravity, electromagnetism, and the strong and weak nuclear forces. We are all familiar with the first two forces on this list. Gravity, among other things, pulls us and other objects downward toward Earth and keeps planets in their orbits around the Sun. The effects of electromagnetism can be seen in static electricity, bar magnets, compasses, and bolts of lightning. These forces are found all around our world in a great many manifestations.

The strong and weak nuclear forces are far less obvious from our perspective. They do not produce many macroscopic effects that we witness in our daily experiences.[1] In fact, their effects are so obscure that they remained completely unknown until the age of quantum physics. It is in the nuclei of atoms where these forces become most evident, and the tapestry of nature's forces becomes more diverse. In the world of atoms, the force of gravity is almost entirely irrelevant—its strength being extraordinarily weak compared to the other three forces. Unlike gravity, however, the electromagnetic force is very relevant to the inner workings of atoms. Whereas gravity holds planets in their orbits around the Sun, it is electromagnetism that pulls electrons toward the center of atoms. The strong nuclear force binds protons and neutrons together into ultracompressed nuclei. The weak force was first discovered as a result of its ability to cause some species of atomic nuclei to decay and emit radiation. As physicists came to learn more about the two nuclear forces, they began to imagine the particles that communicate them. Those force-carrying particles had to be out there somewhere, just waiting to be discovered.

IN 1934, A JAPANESE physicist by the name of Hideki Yukawa was the first to propose the existence of a new type of particle to mediate the effects of the strong nuclear force. In particular, he argued that the strong force could be produced through the existence of yet-undiscovered, massive boson particles, which would

---

[1] That is, unless you count the minor detail of causing the Sun to shine.

later become known as mesons.[2] To account for the observed behavior of the strong force, Yukawa's mesons would have to be
about two hundred times heavier than electrons, but still considerably lighter than either protons or neutrons.

The mass of a meson—or any other type of boson—directly af-
fects the distance over which its corresponding force will be felt.
The reason for this can be traced back to Heisenberg's uncertainty
principle, which holds that quantum particles cannot simultaneously have a precisely defined location and a precisely defined velocity. There is also another aspect of Heisenberg's principle, however,
which pertains to time and energy in the quantum world. The more
that you pinpoint the amount of energy possessed by a particle, the
less you can possibly know about the time at which you determined
the amount of energy. Just as the idea of velocity gets fuzzy and
smeared out over space according to Heisenberg's uncertainty principle, energy also gets smeared out and ill-defined over time.

One consequence of this relationship is that small amounts of energy can be "borrowed" for very short periods of time. Since the amount
of energy in a quantum system is not precisely defined, it is possible for
the quantity of energy to increase for a brief period of time—a quantum
fluctuation. From these fluctuations, entirely new particles can spontaneously be created out of nothing—emerging from the vacuum of
empty space. A tiniest instant of time later, the newly created particles
disappear, becoming annihilated as quickly as they were created. Over
very short periods of time, energy is not conserved in the quantum
world. Nor is the number of particles that are present.

This kind of short-term energy loan is at the heart of how Yu-

---

[2] Yukawa originally named his particle the mesotron, but was "corrected" by
Heisenberg. Heisenberg, whose father was a professor of Greek at the University
of Munich, argued that since the Greek word *mesos* contains no *t* or *r,* meson was
the appropriate spelling.

kawa envisioned the workings of the strong nuclear force. In his theory, loans of this kind are being made constantly and everywhere, each time borrowing enough energy from the vacuum to spontaneously create a meson particle. These particles remain in existence for only a brief fraction of a second—after all, when the energy loan has to be paid back, the particles are destroyed. The heavier the particles are, the more energy has to be borrowed in order to create them, and the sooner the loan has to be repaid.

When photons are created in this way, nature is in no hurry to repay the borrowed energy. As a result, the photons can exist for a long time. This is because photons don't have any mass—none whatsoever—and therefore can be created with an arbitrarily small amount of energy. This is what enables the electromagnetic force to be felt over great distances. Borrowed photons can stay around long enough to make the long trips that are required to communicate the electromagnetic force over a long range.

The strong nuclear force, in contrast, is an extremely short-distance force. It is only felt inside of the nuclei of atoms, which are typically only around a millionth of a billionth of a meter in size. Beyond this range, the effects of the strong nuclear force suddenly become astonishingly weak. Although irrelevant over long distances, at short range, the strong nuclear force is the most powerful force known in our universe.

When Yukawa first proposed the existence of his meson theory of the strong force he was only twenty-seven years old. It took some time, but in 1947, evidence for mesons began to appear among cosmic rays, in much the same way as positrons had been discovered fifteen years before.[3] In 1949, Hideki Yukawa

---

[3] Ten years earlier, in 1937, another type of particle was also discovered in cosmic rays. It had approximately the same mass that Yukawa's meson was predicted to have, and was initially mistaken for this particle. In fact, the particle being detected was a heavy version of the electron, called the muon.

became the first Japanese winner of the Nobel Prize for his meson theory.

Yukawa and other particle physicists were undoubtedly very happy with this impressive accomplishment—the origin of yet another of nature's forces had been uncovered. They had not, however, anticipated what lay beyond this discovery. It turns out that what had been discovered was not simply the meson, as hypothesized by Yukawa, but *one of the many types of mesons*. Beginning in the late 1940s and early 1950s, numerous varieties of mesons and other strongly interacting particles began to show up in particle physics experiments.

AT FIRST GLACE, PARTICLE physics in the 1950s might appear to have been an extremely exciting time of discovery. New particles were constantly being discovered, with no sign of stopping. The first meson, found in 1947, became known as the pion. Shortly after its discovery, machines designed to create and study exotic particles began to be built. These machines use magnets to accelerate charged particles, such as electrons or protons, to great speeds. Once accelerated, the particles are collided into one another, compressing great quantities of energy into a tiny volume of space. To create more massive varieties of particles requires greater quantities of energy—just as Einstein's $E=mc^2$ makes clear. This is precisely what particle accelerators are able to provide.

The construction of the era's most important particle accelerator—the Cosmotron—was completed in 1953. The Cosmotron, located at the Brookhaven National Laboratory on Long Island, New York, was a ring-shaped machine, about seventy-five feet in diameter. Its 288 magnets accelerated particles around in circles to greater and greater speeds. After only one second of acceleration, the protons zipping around the Cosmotron had traveled a whopping 135,000 miles. More importantly, the protons had acquired billions of electron-volts—or giga-electron-volt (GeV)—of energy.[4]

With the development and operation of the Cosmotron and other early particle accelerators, many new types of particles began to appear. The longer the machines searched, the longer the list of discoveries became: three varieties of pions, four types of kaons, along with the delta, the sigma, the lambda, the eta, the xi, and so on and so on. Many of these were types of mesons, similar to what Yukawa had envisioned. Others were particles more similar to protons and neutrons—collectively known as baryons.

Although these many discoveries might seem like an amazing success, it was something of an embarrassment of riches. There was no elegance or underlying structure found in these new mesons and baryons. Their existence did not explain anything and didn't solve any problems. There was no explanation for why these new particles were there to be discovered in the first place.

Particle physicists suddenly found themselves in a situation not very different from the chemists of the nineteenth century. Their field was transforming before their eyes, from a simple and elegant description of matter into a long and complicated laundry list of new particles. There was no understanding of how these particles fit into the structure of our world, or what role they play in our universe's structure. There was no known organizing principle. Particle physics had been reduced to taxonomy.

In the mid-1950s, when asked for the name of one of the newly

---

[4] Particle physicists usually talk about energy and mass in units of electron-volts. An electron-volt is the kinetic energy of an electron after it is accelerated by a one-volt electric field. To create electrons and positrons in the collisions of an accelerator, about a million electron volts of energy are needed. To create pions, hundreds of millions of electron volts are required. With a few billion electron volts of energy (a few GeV), many other kinds of particles can be brought into existence.

discovered particles, the great physicist Enrico Fermi replied, "Young man, if I could remember the names of these particles, I would have been a botanist." Within the space of only a few years, the field of particle physics had devolved from a state of simplicity and economy to a seemingly random expanse of confusion and disarray. If physics were ever to be restored as a simple and elegant description of our world, new ideas—and a new organizing principle—would be needed. Enter symmetry.

SYMMETRY HAS ALWAYS PLAYED a central role in mathematics and physics. Almost every physical law or principle is, in some way, a manifestation of symmetry. Until the twentieth century, however, this connection often went unnoticed and always unappreciated. Before I describe this deep connection between symmetry and nature's laws, it is worth pausing for a moment to consider what physicists and mathematicians mean when they talk about symmetry.

We all have some impression of what symmetry is, and what it means for something to be symmetric. You might think of a vase or a human face as being symmetric, for example. There is a certain balance and evenness that accompanies our perception of these objects. When physicists and mathematicians talk about symmetry, however, they are not using the term as casually as most people do. The mathematical role of symmetry refers to something very precise, and very powerful.

At its most basic level, a mathematical symmetry is nothing more and nothing less than something that remains unchanged. Take a vase, for example. As you rotate a vase around its base, the appearance of its profile doesn't change. The same is not true for a cube, for example, or for some irregularly shaped object that doesn't possess the vase's symmetry. Similarly, when you look at someone's face in a mirror, the reflected image is almost identical, or at least not very different, from

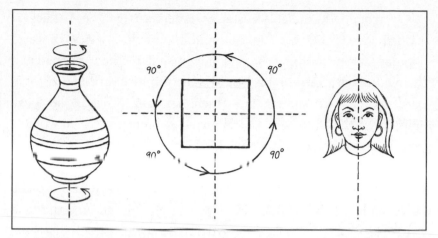

A few examples of geometric symmetry. A vase can be rotated about its axis by any angle without change. A square only remains unchanged if rotated in increments of 90 degrees (90 degrees, 180 degrees, 270 degrees, etc). A human face doesn't possess such a rotational symmetry, but does have an approximate symmetry of reflection (as do the vase and the square).

how it looks normally. This, too, is an example of symmetry— the appearance of a face remains largely unchanged by a mirror reflection. Again, this would not be true for just any object. A person with a patch over his left eye, for example, would appear to have his right eye covered in his reflection; the patch breaks the left-right symmetry.

Symmetry has been an important concept in mathematics for as long as there have been mathematicians. Aristotle, Pythagoras, and other philosopher-mathematicians of ancient Greece considered symmetry to be of paramount importance. Throughout history, countless examples can be found of architecture, art, and music that have the mathematics of symmetry deliberately built into their patterns and structure. But despite this long-standing and rich connection between beauty and symmetry, the role of symmetry in the laws of physics was not really appreciated until comparatively recently, when it was brought to light by a pair of brilliant German mathematicians— Emmy Noether and Hermann Weyl.

Overcoming formidable obstacles, Noether began her studies in mathematics in the year 1900. Despite the birth of a new century, popular sentiments about the role of women in mathematics had not changed much from the backward and primitive attitudes of past ages. In the German city in which her family lived, the local university—the University of Erlangen—entirely prohibited women from being accepted as students. Despite this, Noether managed to gain special permission to sit in on lectures. This exception presumably had something to do with her father, who was a professor of mathematics at the school. Seven years later, she became the first woman to earn a doctorate degree from the university.

Noether did not escape the obstacles faced by female mathematicians upon the completion of her degree. Despite being one of the most talented mathematicians at her university—indeed, in the world—Noether was repeatedly forbidden to apply for academic positions. She ultimately succeeded only by virtue of her undeniable and exceptional talent, made impossible to ignore by the strong and insistent endorsements from several of the world's greatest mathematicians. Despite her exceptional ability and talent, for years she was only allowed to teach without pay. As a further insult, the classes she taught were listed under the name of one her male colleagues, with her official role reduced to that of an assistant. It wasn't until 1923 that she finally began to receive payment for her work.

Among Noether's greatest achievements was something that appears to be a simple—perhaps even obvious—truth to physicists today. This truth was far from obvious at the time, however. What Noether realized was that every conservation law found in nature—such as the conservation of energy, or the conservation of electric charge—is connected in an intimate way to a symmetry. According to what is now known as Noether's theorem, symmetry and conservation laws imply and require each other. They are, indeed, unavoidable consequences of each other.

Take, for example, the law of conservation of energy. This simple principle states that the quantity of energy in any given isolated system never changes. The amount of energy present now is the same that there was at all times in the past, and the same as there will be at every point of time in the future. This is no different from stating that the laws of physics do not change with time; Noether's theorem equates the law of conservation of energy with a symmetry in time. Similarly, the fact that (as far as we know) the laws of nature are the same at all points in space is a symmetry of space, corresponding to the law of conservation of momentum.

At around the same time that Emmy Noether proved her famous theorem, her friend and colleague Hermann Weyl was studying another kind of symmetry. Weyl recognized that there were interesting properties of symmetry dealing with the way in which physical properties could be measured. He named this mathematical property "gauge symmetry."

Trying to explain the concept of gauge symmetry can be very tricky. To help understand what these symmetries are all about, let's consider a couple of examples. First, imagine that I am climbing Mount Everest. I begin from the base camp, at an elevation of 17,500 feet, and climb to the summit at 29,028 feet. These altitudes are each measured in the standard way, relative to sea level. If the oceans' water wasn't there, we would probably instead define elevation relative to the lowest point on Earth—what is now the deepest location at the bottom of the ocean, Challenger Deep, in the Mariana Trench. From this reference point, the Everest base camp is at an elevation of 53,338, and the summit is at 64,866 feet. Regardless of which elevation scheme I use, however, the *change* in elevation that I experience as I make my climb is the same. In either case, I have ascended 11,528 feet.

Similarly, lines of longitude are generally defined relative to the prime meridian. If this line of zero degrees longitude were defined elsewhere, nothing would change—meaning that the change of longitude between two places on the globe would be the same. For the purposes of navigation or any other practical consideration, it doesn't matter whether the prime meridian goes through Greenwich, England, or New York City, or Cold Spring, Minnesota, or any other point on Earth.

Electric voltage has the same property. Voltage itself is an arbitrary quantity—only differences in voltage really matter. A bird can safely land on a high-voltage wire because it is not in contact with a low-voltage object—it is not grounded. Although differences in voltage can be deadly, absolute voltages are harmless—indeed, meaningless. Voltage, just like elevation and longitude, is defined by an arbitrary scale. Only *differences* in voltage, elevation, and longitude are physically relevant.

Each of these is an example of a global gauge symmetry. Weyl realized, however, that there are also other kinds of gauge symmetry. We say that a symmetry is global if it relies on all of the locations or voltages or whatever being shifted together. We can add the same thirty degrees to the longitude of every city in an atlas without changing anything, but we cannot add thirty degrees to London and fifteen to New York without altering measurable quantities—namely, the difference in longitude between the two cities. Similarly, if I redefined the voltage scale at the point of ground beneath your left foot, but not at the point beneath your right foot, it could very well lead to your electrocution. With global gauge symmetries, the arbitrary scale can be changed without affecting anything, but it must be changed at every location at the same time.

Not all gauge symmetries are global, however. There are also local gauge symmetries, which Weyl found especially intriguing. Local gauge symmetries derive their power not from the relationships between different points in space, as global symmetries do, but at each point in space independently of one another. In a sense, local gauge symmetry is something like having a countless number of global symmetries, at every single point in space.

Weyl knew that the equations of electromagnetism possess a local gauge symmetry. This symmetry basically comes down to the arbitrariness of a quantum property called the phase of a field. You can arbitrarily change the quantum phase at every point in space freely without really changing anything. The mathematical structure of the theory guarantees that any change induced in the quantum phase will automatically be counteracted and corrected for by the other fields.

Although no one could see it fully yet, Weyl was beginning to get a glimpse at just how spectacularly powerful local gauge symmetry could be. It is, indeed, at the very heart of our understanding of the quantum world.

THE REAL STRENGTH AND importance of Emmy Noether's and Hermann Weyl's insights did not become evident until half a century later. Although Weyl had known that electromagnetism possessed local gauge symmetry, he and others weren't sure whether this was of any great significance. It certainly didn't appear to be useful or predictive. In the 1950s, however, many new types of particles—various species of mesons and baryons—were discovered, which in turn revealed the presence of many new conservation laws. According to Noether's theorem, these new conservation laws must imply new symmetries. We now understand that these conservation laws correspond to local gauge symmetries, much like the symmetry Weyl had tied to electromagnetism. This connection—beginning with Noether and Weyl—provides the very foundation of modern particle physics.

The new conservation laws that were discovered in the 1950s are somewhat different from those you may have learned about in high school physics. The quantities that are conserved according to these laws are not familiar things like energy or momentum. Instead, they apply to quantities known as quantum numbers. Every type of particle has its own set of quantum numbers. Electric charge is one kind of quantum number, for example. Electrons have −1 electric charge, whereas positrons and protons each have +1. Photons have no electric charge. The total amount of electric charge never changes in any known interaction or process. If a process creates or destroys a negatively charged particle such as an electron, for example, it must also create or destroy a positively charged particle to balance the change in electric charge. As far as we know, the conservation of electric charge is a law that is absolutely never broken.

By the end of the 1950s, a total of six different quantum numbers had been discovered. These six quantities—electric charge, spin, isospin, parity, baryon number, and strangeness—are found in various combinations in different mesons and baryons. It is these quantum numbers that dictate what kinds of interactions each type of particle can undergo. All of the observed interactions conserve the total amount of electric charge and baryon number, for example. Some of the other quantum numbers, such as isospin and strangeness, however, are conserved only in certain types of interactions, and can change in others. Parity is a quantity that has to do with the orientation of a particle—to reverse a particle's parity would be like looking at it in a mirror. Spin is a quantity that tells us how much angular momentum a particle can have. It is also this property that determines whether a particle is a boson or a fermion—the difference between being simple matter or a carrier of force. Any particle with a spin of one or zero, or any other integer, is a boson. Fermions, in contrast, have half-integer spins, like one-half, one and a half, and so on.

The particles that had been discovered by the 1950s had many different combinations of the various quantum numbers (see Table 1).

PARTICLE NAME	SPIN	ELECTRIC CHARGE	ISOSPIN	BARYON NUMBER	PARITY	STRANGENESS	MASS
Proton (p)	1/2	+1	+1/2	+1	+1	0	0.938 GeV
Neutron (n)	1/2	0	−1/2	+1	+1	0	0.940 GeV
Neutral sigma (Σ⁰)	1/2	0	0	+1	+1	−1	1.19 GeV
Negative sigma (Σ⁻)	1/2	−1	−1	+1	+1	−1	1.20 GeV
Positive sigma (Σ⁺)	1/2	+1	+1	+1	+1	−1	1.19 GeV
Lambda (Λ)	1/2	0	0	+1	+1	−1	1.12 GeV
Neutral xi (Ξ⁰)	1/2	0	+1/2	+1	+1	−2	1.32 GeV
Charged xi (Ξ⁻)	1/2	−1	−1/2	+1	+1	−2	1.32 GeV
Negative delta (Δ⁻)	3/2	−1	−3/2	+1	+1	0	1.23 GeV
Neutral delta (Δ⁰)	3/2	0	−1/2	+1	+1	0	1.23 GeV
Positive delta (Δ⁺)	3/2	+1	+1/2	+1	+1	0	1.23 GeV
Doubly positive delta (Δ⁺⁺)	3/2	+2	+3/2	+1	+1	0	1.23 GeV
Neutral sigma star (Σ*⁰)	3/2	0	0	+1	+1	−1	1.39 GeV
Negative sigma star (Σ*⁻)	3/2	−1	−1	+1	+1	−1	1.39 GeV
Positive sigma star (Σ*⁺)	3/2	+1	+1	+1	+1	−1	1.39 GeV
Neutral xi star (Ξ*⁰)	3/2	0	+1/2	+1	+1	−2	1.53 GeV
Charged xi star (Ξ*⁻)	3/2	−1	−1/2	+1	+1	−2	1.53 GeV
Neutral kaon (K⁰)	0	0	−1/2	0	−1	+1	0.50 GeV
Charged kaon (K⁺)	0	+1	+1/2	0	−1	+1	0.49 GeV
Neutral pion (π⁰)	0	0	0	0	−1	0	0.13 GeV
Charged pion (π⁺)	0	+1	+1	0	−1	0	0.14 GeV
Eta (η)	0	0	0	0	−1	0	0.55 GeV

Table 1. The list of mesons and baryons known as of 1962 (not including their antiparticles), and the values of each of their quantum numbers and mass.

The situation these discoveries represented is taxonomy at its very worst. Physicists wanted to know why there were particles with these particular properties. Why did particles exist with strangeness of −1 and an electric charge of +1 (the positive sigma), and with strangeness of 0 and electric charge of +2 (the doubly charged delta), but not with −1 electric charge and strangeness of +2? Furthermore, why were there six different quantum numbers, and what do they have to do with one another? Was there an order behind this chaos? And if so, what could that order possibly be?

THE PATH TOWARD RESTORING order and structure into our understanding of the quantum world followed directly from the ideas of Emmy Noether and Hermann Weyl. The six quantum numbers known in the late 1950s are nothing more than conservation laws. Noether's theorem tells us that conservation laws correspond to the existence of a symmetry. The symmetry these quantum rules collectively correspond to is, in fact, a local gauge symmetry, as had been studied by Weyl.

Although other physicists could see the trailhead of the path set out by Noether and Weyl, it took the genius of Murray Gell-Mann to navigate the difficult terrain of the route. Over the course of a few years, Gell-Mann took the cluttered mess of particles discovered in the 1950s and converted them into a simple and ordered structure. Not much later, he even explained where that order came from. His insights were one of the greatest scientific achievements of all time.

Murray Gell-Mann's exceptional talents were evident from a very young age. At the age of three, he was able to multiply large numbers in his head and had learned to read. Barely out of infancy, he found himself studying Roman history, language—Latin, French, Spanish, and even Egyptian hieroglyphics—music, and mathematics. He found himself winning spelling bees and other academic

Murray Gell-Mann and Richard Feynman

contests, often against much older children. Despite being the son of far-from-wealthy immigrants in New York City, the young Murray's talents were too spectacular to go unnoticed. At the age of eight, he left his local elementary school to enter the prestigious Columbia Grammar and Preparatory School, where he was given a full scholarship. At the age of fifteen, he was accepted for study at Yale University. After earning degrees from Yale and MIT, he went on to pursue research at Princeton, the University of Chicago, and Caltech.

While at Chicago and Caltech, Gell-Mann devoted his attention to the newly discovered mesons and baryons, hoping to find some order behind the complicated mess. He began by considering the aspects of particle physics that were already fairly well understood. In particular, he knew that the symmetries behind the theory of quantum electrodynamics—as established by Paul Dirac and others—could be described by a mathematical structure called the U(1) symmetry group. Gell-Mann began to wonder whether there might be another symmetry group that described the behavior of the new mesons and baryons.

Mathematically speaking, the U(1) group describes the most simple of all possible local gauge symmetries. In terms of particle physics, it corresponds to a single type of boson (the photon) that acts on a single kind of quantum number (electric charge). To encompass the complex and diverse aspects of the many newly discovered quantum particles, Gell-Mann knew that he would need a much larger and more complex gauge symmetry. Unfortunately, few physicists in the 1950s knew much about the mathematics of symmetry—called group theory. Fortunately, Murray Gell-Mann was not an ordinary physicist.

The complexity of a given type of symmetry group can be described by what is known as its number of generators. In the case of the U(1) group of quantum electrodynamics, there is only one generator—associated with electric charge. Other theoretical physicists had also been working with a larger group, called SU(2), which has two generators. Neither of these groups is even close to being capable of encompassing the many new quantum numbers possessed by the mesons and baryons, however. Gell-Mann gradually considered larger and larger groups—with three, four, five, six, and even seven generators—determined to find one that could encompass all of the known particles and symmetries. None of these worked. Months passed without significant progress.

The symmetry groups Gell-Mann had been considering were each combinations of U(1) and SU(2) groups. In fact, all gauge groups with seven or fewer generators can be built up from combinations of these two group types. In the fall of 1960, however, Gell-Mann learned about another group, called SU(3). Unlike the others he had been considering, SU(3) wasn't a simple combination of U(1) and SU(2). Instead, it was the smallest possible gauge symmetry that is independent of U(1) and SU(2). Almost immediately, Gell-Mann recognized that this looked like the symmetry he had been searching for.

The eight generators of the SU(3) group bring forth a very rich and interesting mathematical structure. Two of the generators correspond to the strangeness and isospin quantum numbers—just as electric charge corresponds to the lone generator of the U(1) group. The other six of the SU(3) generators are rules for changing the quantum numbers' values: isospin +1 and strangeness 0, isospin −1 and strangeness 0, isospin +1/2 and strangeness +1, isospin −1/2 and strangeness +1, isospin +1/2 and strangeness −1, and isospin −1/2 and strangeness −1.

The next step was for Gell-Mann to build a table of possible particle species from the rules of the SU(3) group. If SU(3) really was the symmetry behind the new particles, then the table should be able to predict their properties. If not, he would have to start all over again with another symmetry group.

The sets of particles that come out of a symmetry group are called representations. The simplest possible representation of the SU(3) group contains only three members. Since this is far too few to explain all of the observed varieties, Gell-Mann quickly skipped it over and went straight to the next largest choice. Following the rules of the group, he wrote down all of the combinations of strangeness and isospin that were allowed. To his amazement, these combinations exactly match the properties of eight of the lightest baryons—the proton, the neutron, the three sigmas (Σ^+, Σ^0, and Σ^-), the lambda (Λ), and the positive and negative xi (Ξ^+ and Ξ^-). Furthermore, these eight particles all had the same value for their spin (1/2) and parity (+1). He called this the "baryon octet." Seven of the known mesons with the same spin and parity also fit nicely into an octet. As an irreverent gesture to the success of his octets, he called his new scheme the "eightfold way," after the

THE BARYON OCTET

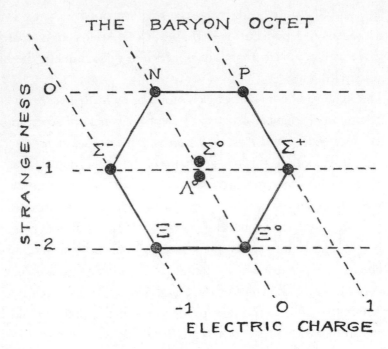

THE MESON OCTET

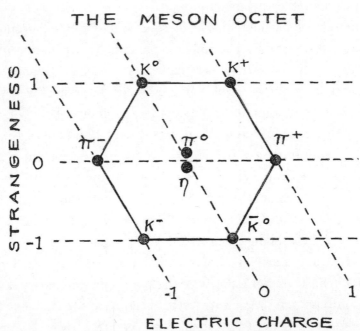

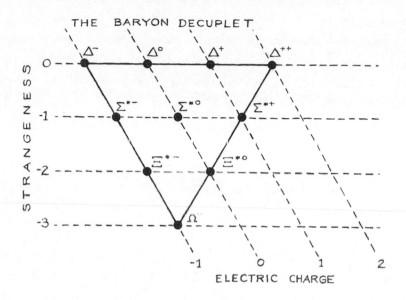

Murray Gell-Mann's baryon octet, meson octet, and baryon decuplet, as predicted by the SU(3) symmetry group. These patterns provide an explanation for why the mesons and baryons are found with their various combinations of electric charge, strangeness, and other quantum numbers.

eight aspects of the path to enlightenment, as professed by the Buddha.[5]

Moving on from the octets, Gell-Mann also worked out the next largest representation of SU(3)—a decuplet of ten baryons. Of these ten particles, only four corresponded to particles that had been discovered. If Gell-Mann was right, then there must be six other baryons that hadn't been discovered yet. Furthermore, one meson was missing from his octet—an electrically neutral meson, with no isospin and no strangeness, which he called the eta (η).

---

[5] This remark, although intended to be nothing but humorous, has been used by many to support the idea that there is some kind of relationship between quantum physics and Eastern mysticism. This seems to have irritated Gell-Mann to no end.

Compelled by the mathematics, he predicted the existence of the missing baryons and meson. He even estimated how massive the undiscovered particles should be. Over the following few years, one by one, Gell-Mann's particles were discovered. In 1964, the last of these—the omega minus (Ω^-)—was discovered at the Brookhaven National Laboratory. The omega minus had precisely the properties that had been predicted by Gell-Mann. With this final confirmation, the eightfold way was universally accepted as the organizing principle behind the many mesons and baryons. Gell-Mann's SU(3) symmetry is in fact built into the very blueprint of our universe.

EVEN IF GELL-MANN HAD stopped at this point and never done another piece of original physics research, he would still have gone down in history as one of the most important figures in twentieth-century physics and would almost certainly have earned the Nobel Prize. The development of the eightfold way had secured him a revered place in the history of science.

Gell-Mann's research did not end with the eightfold way, however. He had discovered that SU(3) symmetry fits into our world, but was not content with this knowledge. Looking further ahead, he wanted to understand *why* this symmetry that he had discovered was present in the structure of the mesons and baryons. Where did it come from?

In 1869, Dmitry Mendeleyev introduced the periodic table of the chemical elements that we all learned about in high school chemistry. Today, we know that the entries found in that table are each combinations of protons, neutrons, and electrons. In Mendeleyev's time, however, no one knew about these smaller constituents. But despite this, Mendeleyev was able to recognize the patterns in the periodic table, even without understanding what lay beneath it. Gell-Mann had done something similar to this with the mesons and the baryons. He recognized the patterns of SU(3), but didn't know where it came from, or what lay beneath.

By studying the patterns in his periodic table, Mendeleyev was able to predict the existence of new chemical elements—including germanium, gallium, and scandium—before they had been discovered. In much the same way, Gell-Mann was able to predict the existence of the omega minus and other particles from the patterns of SU(3). In 1907, Mendeleyev's life came to an end before he ever understood why his table worked or where it came from. Only after the discovery of the proton in 1919 and the neutron in 1932 did the underlying logic behind the periodic table become clear. Unlike Mendeleyev, Gell-Mann lived to understand the mysteries behind his symmetry—in fact, he was the one to solve this puzzle. Barely pausing to take a breath following the experimental confirmation of the eightfold way, he turned to the next step in exploring his symmetry. What he found behind the symmetry of SU(3) were not the protons, neutrons, or electrons of the periodic table, but something weirder—quarks.

Gell-Mann was never content with his decision to ignore the smallest possible representation of the SU(3) symmetry group—the one with only three members. As there are far more than three mesons and baryons, he had felt forced to discard it and move on to larger representations. Nevertheless, the choice haunted him. Somehow, he thought, the smallest and most fundamental representation of the symmetry must be manifest in nature. Everything else must come forth from it—but how?

In 1964, following the discovery of the omega minus, Gell-Mann returned to the smallest representation of SU(3). Considering the mathematics behind this symmetry and how it might be realized in nature, he began to imagine that all of the types of mesons and baryons must be made up of a smaller number of particles, as revealed by the fundamental representation of SU(3). In a quirk of dry humor—reminiscent of his "eightfold way"—he named these fundamental particles "quarks," after a line from James Joyce's stream- of-consciousness novel *Finnegan's Wake:* "Three quarks for Muster Mark!"

Gell-Mann postulated that quarks come in three varieties—which he called up, down, and strange quarks. Each of these quark types has a set amount of electric charge, strangeness, and other quantum properties. By combining these three species of quarks in different ways, all mesons and baryons could be constructed. Each meson, Gell-Mann realized, could consist of a quark and an antiquark bound together, and each baryon of three quarks, and antibaryons of three antiquarks. In Table 2, the amount of electric charge and strangeness carried by each of the three quark types is given. These three simple quarks can be combined in groups to produce every single one of the baryons and mesons. Finally, Gell-Mann had uncovered the simple origin of the numerous mesons and baryons.

When Gell-Mann first proposed his quark hypothesis, a very serious objection was raised. Namely, Gell-Mann's scheme appeared to violate the Pauli exclusion principle. You might remember from chapter 3 that Wolfgang Pauli had shown, decades before Gell-Mann proposed his quark model, that no two fermions could ever be in precisely the same quantum state at the same time. This was the idea that Dirac used to argue electrons wouldn't fall into the negative-energy sea because those states were already full. Well, in Gell-Mann's theory, many of the baryons consist of quarks bound up together in exactly the same state. A proton is three quarks—two up quarks and

QUARK TYPE	ELECTRIC CHARGE	STRANGENESS
Up (u)	+2/3	0
Down (d)	−1/3	0
Strange (s)	−1/3	−1

Table 2. The three quark types proposed by Murray Gell-Mann. The three species of antiquarks have opposite values for their electric charge and strangeness.

PARTICLE NAME	ELECTRIC CHARGE	STRANGENESS	QUARK CONTENT
Proton (p)	+1	0	uud
Neutron (n)	0	0	udd
Neutral sigma (Σ^0)	0	−1	uds
Negative sigma (Σ^-)	−1	−1	dds
Positive sigma (Σ^+)	+1	−1	uus
Lambda (Λ)	0	−1	uds
Neutral xi (Ξ^0)	0	−2	uss
Charged xi (Ξ^-)	−1	−2	dss
Negative delta (Δ^-)	−1	0	ddd
Neutral delta (Δ^0)	0	0	udd
Positive delta (Δ^+)	+1	0	uud
Doubly positive delta (Δ^{++})	+2	0	uuu
Neutral sigma star (Σ^{*0})	0	−1	uds
Negative sigma star (Σ^{*-})	−1	−1	dds
Positive sigma star (Σ^{*+})	+1	−1	uus
Neutral xi star (Ξ^{*0})	0	−2	uss
Charged xi star (Ξ^{*-})	−1	−2	dss
Omega minus Ω^-	−1	−3	sss
Neutral kaon (K^0)	0	+1, −1	d anti-s, s anti-d
Positive kaon (K^+)	+1	+1	u anti-s
Negative kaon (K^-)	−1	−1	s anti-u
Neutral pion (π^0)	0	0	u anti-u, d anti-d
Positive pion (π^+)	+1	0	u anti-d
Negative pion (π^-)	−1	0	d anti u
Eta (η)	0	0	u anti-u, d anti-d, s anti-s

Table 3. The quark content of the mesons and baryons; u, d, and s denote up, down, and strange quarks, respectively.

a down quark—bound together. The two up quarks in a proton are in the same quantum state. Similarly, the two down quarks in a neutron are in the same quantum state. Quarks are fermions and have to follow the exclusion principle. According to this reasoning, quarks simply couldn't be combined in this way.

There is, however, a way that particles can get around the stringent rules of Pauli's principle. In particular, Pauli's rule only applies to identical particles. If there was some quantum property that made each of the quarks in a baryon different from one another, then they could be bound up together without any problem. So, in order to evade the objections that had been raised about his quark model, Gell-Mann hypothesized that quarks must carry yet another type of quantum number. So, just as electrons, positrons, and protons carry electric charge, quarks carry another type of charge called color. But whereas electric charge comes in only one variety, color comes in three—labeled red, blue, and green. Each quark is either red, blue, or green. Each antiquark is antired, antiblue, or antigreen—anticolors being something analogous to negative electric charge.

Gell-Mann called this new quantum charge color because it shared—by analogy—a well-known property of light. When all colors of light are combined, you are left with white light—colorless.[6] Similarly, when one red, one green, and one blue quark are combined, the total color of the baryon is zero. Additionally, when a quark of a given color is combined with an antiquark with opposite color, the total color of the resulting meson is also zero.

There is, however, one rather awkward aspect of the quark model of baryons and mesons—we have never seen a quark isolated in nature. Experiments have revealed groups of quarks bound up together in mesons and baryons, but never a quark by itself. This is like observ-

---

[6] Beyond this analogy, Gell-Mann's "color" has nothing to do with the colors seen in light. A blue quark would resemble the hue of the sky no more and no less than a red or a green quark.

ing the atoms of the periodic table, but never detecting a lone proton, neutron, or electron. So what keeps us from discovering lone quarks?

The answer to this question is that quarks are absolutely bound together and confined inside mesons and baryons. You can try all you want to tear apart a meson or a baryon into its constituent quarks, but you will never succeed. The more the quarks become separated, the more they pull themselves together—like a rubber band that becomes harder to pull the longer it is stretched. This makes the force holding quarks together—the strong force—very different from the other forces of nature. The electromagnetic force holds electrons in orbits around atoms, but, with enough energy, those electrons can be freed. Similarly, the force of gravity confines us to Earth's surface. Yet despite gravity's pull, it is frequently overcome by rockets. If you were to throw a baseball upward with a speed greater than about twenty-five thousand miles per hour, it would travel away indefinitely, never falling back to Earth. This is called Earth's escape velocity. But unlike gravity, the strong force has no escape velocity. The more you pull apart, the more the strong force resists.

If you look at the list of all possible mesons and baryons, you will notice a peculiar feature—despite being made up of quarks that each carry color, all of the particles that can exist isolated in nature have exactly zero net color. Every meson contains one quark of a given color and an antiquark of the opposite anticolor, which together constitutes no color. Each baryon contains three quarks—one of each color, for a total of zero color. No isolated colored object has ever been found in nature. Color is something that exists among quarks contained inside baryons and mesons, but cannot exist independently.

QUARKS ARE ONLY ONE side of Gell-Mann's SU(3) theory. In the quantum theory of electromagnetism, there are charged fermion particles (such as electrons) and then there are the boson particles that mediate and communicate the electromagnetic force (photons). In

Yukawa's original theory of the strong force, it was the mesons that communicated the strong force. With the development of the SU(3) theory behind mesons, baryons, and the strong force, the true underlying origin of this force became evident. In Gell-Mann's theory, in addition to the colored fermions—the quarks—there are also bosons that possess color. These manifestations of the strong force are called gluons.

The number of bosons that a gauge symmetry—such as SU(3)—contains depends on its number of generators. The gauge symmetry behind electromagnetism—the U(1) symmetry—has only one generator, and therefore contains only one variety of boson: the photon. SU(3) is a far more complex group, containing eight generators and, therefore, eight types of gluons. These eight gluons are very different from the photons of electromagnetism. For one thing, although photons communicate the force between electrically charged particles, they do not themselves possess any electric charge. Gluons, on the other hand, are themselves colored particles. Each gluon possesses one color *and* one anticolor—red-antiblue, or green-antired, for example.[7] As a result, whenever a quark comes into contact with a gluon, its color changes. For example, a red quark could radiate a red-antiblue gluon, which would afterward leave the quark blue. In this process, the gluon carries away the redness of the quark, and leaves behind its blueness.

Just like quarks, individual gluons are never seen in isolation. They are found together with quarks inside of mesons and baryons, but never alone. There is reason to think that groups of gluons should be able to exist in groups without quarks—called "glueballs"—but such states are yet to be discovered.

---

[7] Since there are nine possible combinations of color and anticolor, you might think that this should lead to there being nine types of gluons instead of eight. For technical reasons having to do with the structure of the SU(3) symmetry, however, this is not the case.

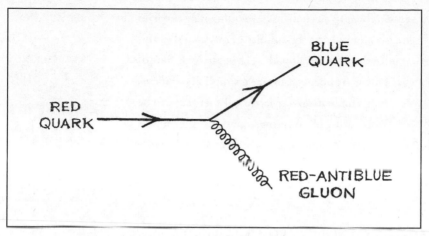

An example of an interaction between quarks and gluons.

IN THE COURSE OF only a few years, Murray Gell-Mann had taken a chaotic mess of particles—lacking any semblance of order or pattern—and discovered the underlying structure that lay beneath. He had not only discovered the patterns and made predictions based on them; he also revealed the reasons *why* those patterns were there to be found in the first place. With amazing insight and mathematical prowess, he took the science of physics from chaos to order, and from frustration to beauty.

Perhaps even more important, with these discoveries, Gell-Mann demonstrated the central importance of symmetry in understanding the underlying rules that govern our universe. The mathematics of symmetry enable us to determine the possible forms these rules can take, and guide us in our search to understand the structure our world is based upon.

Symmetry is not only useful. It is essential.

IIIIIIIIII **5** IIIIIIIIII

The World as We Know It

With the discovery of quarks, gluons, and the eightfold way, Murray Gell-Mann had opened the door to what symmetry could bring to the science of particle physics. But even after this great accomplishment, the door of symmetry's promise was only opened a crack. The potential power and insight that lay beyond that passage was still far from being fully appreciated.

In 1964, with the experimental confirmation of the SU(3) symmetry in the form of the discovery of the omega minus particle, a greater understanding of our world was well on its way to being constructed. The strong and the electromagnetic forces had become understood entirely within the context of gauge symmetries. The picture of particles and force that had emerged was remarkably elegant and simple. More important, the theory behind this picture was very powerful and predictive. But that was only the beginning.

By the early 1960s, much had become understood about the

underlying structure of our universe, but there were still many questions without answers. For one thing, there was no satisfactory theory to describe the weak force, and lingering behind each discovery was the question, what is it about gauge symmetry that leads to its connection with particle physics? Much remained to be learned.

Gradually, answers were found to these questions and others. When all was put together, what resulted was the framework of particle physics that goes by the staggeringly understated name "Standard Model." But don't let its name fool you. The Standard Model is perhaps the greatest—and most rigorously confirmed—achievement in the history of science.

The way that words such as *model, law,* and *theory* are used in science can be very confusing and even frustrating. Despite the all-too-common perception to the contrary, these labels are attached to scientific ideas for reasons that have nothing to do with how confident scientists are in their validity. Compare, for example, Einstein's theory of relativity and Newton's law of gravitation. Based on the common usage of the words *theory* and *law,* one might jump to the conclusion that we know—or are at least fairly confident—that Newton's description of gravity is correct and that Einstein's description may or may not be. Relativity is, after all, "just a theory." But that is not at all what scientists mean when they use these words. Einstein's theory has been tested with amazing precision, and has never been shown to be wrong by any experiment whatsoever. Newton's law of gravitation works quite well, but only under certain conditions. The planet Mercury, for example, is so close to the Sun that the force of gravity cannot be accurately described by Newton's equations. To understand why Mercury orbits the Sun the way it does, Einstein's theory is required. Einstein's theory is more accurate and more widely applicable than Newton's law.

The most frequent and egregious abuses of the "theory-law" language are made by those attempting to discredit Charles Darwin's theory of evolution. As with Einstein's relativity, evolution being "just a theory" has nothing to do with its validity. Biological evolution through the mechanism of natural selection has been repeatedly tested and confirmed. Virtually no professional biologists challenge this. Being a "theory" has nothing to do with the degree of confidence the scientific community has in it.

Much like the theories of Einstein and Darwin, the Standard Model also possesses a potentially misleading name. Since its invention in the early 1970s, the predictions of this theory have been tested rigorously, with ever increasing precision. Many scientists expected this model to eventually break down and fail to describe accurately the phenomena observed in our universe. But despite these expectations, it has firmly stood the test of time. The degree of experimental success of the Standard Model is as great as any other scientific law or theory that has ever been. The amazing range of accuracy and applicability of this theory is rivaled only perhaps by its exquisite elegance and simplicity.

JUST AS GELL-MANN'S EIGHTFOLD way is based upon the principle of gauge symmetry, so is the Standard Model. In the language of group theory, the structure behind the Standard Model can be written as $SU(3) \times SU(2) \times U(1)$. The $SU(3)$ part of this expression describes the eightfold way and the interactions of the strong force as discovered by Gell-Mann. The remaining part of the symmetry group—$SU(2) \times U(1)$—represents the combined nature of the electromagnetic and weak forces. Within the context of the Standard Model, these two forces are not independent and separate entities. They are instead dual aspects of one interaction, known as the electroweak interaction. The electromagnetic and weak forces are two seemingly very different sides of the same underlying coin.

FORCE	BOSON(S)	ACTS UPON	MASS
Strong	8 gluons	Colored particles	0
Electromagnetic	Photon	Electrically charged particles	0
Weak	Z, W⁺, W⁻	All known fermions	91.2 GeV, 80.4 GeV

Table 4. The forces and corresponding bosons of the Standard Model.

The forces of the Standard Model are brought into existence by a number of different types of boson particles—eight gluons, the photon, and three communicators of the weak force, called simply the Z, W⁺, and W⁻ bosons. The properties and characteristics of these bosons are what determine the way in which the forces act—their strength, their range, and what they act upon.

When Hideki Yukawa first formulated his early model of the strong force, he concluded from the short range of the force that it must be communicated by a heavy boson—what he called a meson. We now understand that Yukawa's theory was missing many of the elements needed to adequately describe the interactions of the strong force. At the time, he didn't know about quarks and gluons or how they are confined together into mesons and baryons. As it turns out, the bosons that mediate the strong interaction—gluons—don't have any mass at all. Yukawa's reasoning, although insightful, didn't apply very well to the complicated dynamics behind the nature of the strong force.

For another kind of force without these complications, however, Yukawa's relationship between a force's range and the corresponding boson's mass can be successfully applied. The heavier a boson is, the shorter the range of its resulting force will be. This connection can be seen clearly in the behavior of the weak force.

Much like the strong force, the weak force is felt only over very short distances—even shorter distances than the strong force. For this reason, these two forces are sometimes called "nuclear forces," since their effects were originally observed only inside the nuclei of atoms, where particles are packed in very close to one another.[1] Much like in Yukawa's meson theory, the reason that the range of the weak force is so short is that its mediating bosons are very, very heavy. When these particles—the W^+, W^-, and Z bosons—were first proposed in the late 1960s, the heaviest known particle was the omega minus—the baryon made up of three strange quarks—which weighs about 1.7 GeV.[2] The W and Z gauge bosons, in contrast, weigh a whopping 80.4 and 91.2 GeV, respectively. The heavier a particle is, the more difficult it is for it to be produced and studied in particle collider experiments. With the technology and experiments that existed in the 1960s and 1970s, there was no way that such incredibly heavy particles could be observed. Such particles must have seemed like pure philosophy at the time—a hypothesis with no tangible predictions that could be put to the test. But one should be wary of underestimating the rate of science's progress. In 1983, these particles were observed directly for the first time. Throughout the 1990s, millions of W and Z bosons were produced and studied by an experiment called LEP—short for the Large Electron-Positron Collider.

The short range of the weak force is not the only effect of the heaviness of the W and Z bosons. The very large masses of these particles also explains why the weak force is, well, so weak. Whereas the electromagnetic force is able to bind electrons to atomic nuclei, and the strong force holds together those nuclei themselves, the

---

[1] In modern experiments, we can see the effects of the strong and the weak forces in environments other than atomic nuclei, however, making the "nuclear" label somewhat antiquated.

[2] Once again, I am talking about quantities of mass in terms of energy, as related by $E = mc^2$. From now on, I will describe masses and energies each in terms of GeV (giga-electron-volts).

weak force binds nothing. Particles feel the influence of the weak force in only the mildest of ways.

To give you an idea of how feeble the weak force is, consider for a moment some of the ways that the three forces of the Standard Model are manifest. For example, these three forces can each cause certain types of particles to decay, disintegrating into smaller and lighter particles. When the electromagnetic force is at work, it typically takes something like a billionth of a billionth of a second for the decay to take place. Particles that can decay through the effects of the strong force disintegrate even more quickly. The weak force, however, being so feeble, takes much longer to act. Consider the lifetimes of neutral and charged pions, for example. Neutral pions decay through the electromagnetic force, and survive for only about 10^{-16} seconds, on average, before disintegrating. Charged pions, in contrast, only decay through the effects of the weak force, and live for about 10^{-8} seconds. Although this is still a very short time—ten-billionths of a second or so—it is very much slower than the decay times of the other forces. The weak force takes millions of times longer because it is millions of times weaker.

When quarks are bound together by the strong force into mesons and baryons, they cannot be pulled apart. No isolated quark can exist without others around it. The strong force is simply too powerful. The electromagnetic force is what holds electrons in their orbits around atoms—much like gravity holds the planets in their orbits around the Sun. Although the electromagnetic force can be overcome with enough energy, it is usually sufficient to maintain an electron's orbit. In contrast, the weak force cannot hold any two particles together. The gentle tug of the weak force is far too feeble to bind. It is as if it is only able to offer a mere suggestion, while the electromagnetic force issues demands and the strong force speaks with the bark of a drill sergeant's command.

I mentioned before that according to the Standard Model, the electromagnetic force and the weak force are inseparable aspects of the same phenomena, called the electroweak force. One aspect of this connection is seen in the fact that the bosons associated with these forces—the photon for electromagnetism and the W and Z bosons for the weak interactions—communicate with the particles they act upon with similar intensities. A particle physicist would say they have similar "couplings." But if the couplings are so similar, who do these two forces appear to us to have such very different strengths? Two electrons of modest energy exchanging a photon will repel each other with much more strength—thousands, millions, or even billions of times more—than can be generated by the exchange of a Z boson. The couplings are similar, but the forces are very different. The reason for this can be traced back once again to Heisenberg's uncertainty principle.

When the two electrons interact electromagnetically by exchanging a photon, there is plenty of energy available to use to create that photon and move it through space to the other electron. Photons, after all, have no mass and require almost no energy to create. In order for the same two electrons to exchange a Z boson of the weak force, however, a lot of energy is required. A Z boson has a mass equivalent to 91.2 GeV of energy—much, much more than is possessed by two typical electrons. In order to create a Z boson to exchange, thus bringing the weak force into being, 91.2 GeV of energy has to come from somewhere—to be "borrowed" from the vacuum of empty space. According to Heisenberg's uncertainty principle, quantities of energy can fluctuate up and down, sometimes even enough to spontaneously create a Z boson. The more energy it takes to create a particle, however, the more rare such spontaneous events are. Z bosons are created from the nothingness of empty space, but less frequently and for shorter periods of time than photons. Because they are far scarcer than photons, the force the Z bosons generate is far weaker than electromagnetism.

Imagine two investors, each of whom has nothing but a hundred dollars in his pocket. They each learn about an investment opportunity. The first gets a tip about an incredible stock that he can buy for a hundred dollars and get a million dollars back the next day. The second investor learns of another block of stock for sale for a million dollars that will be worth $2 million tomorrow. Of course both investors want to cash in. For the first investor, this is easy. He just buys the stock for the hundred dollars in his pocket and waits for the profits to come in. The second investor somehow has to borrow a million dollars before he can participate. If he manages to find a lender, both investors will make roughly the same amount of money. If he doesn't, then the first investor will get rich while the second gets nothing. Similarly, the electromagnetic and weak forces have very similar couplings, but, to use those couplings, the weak force needs to borrow an enormous amount of energy. But nature, like a cautious banker, is hesitant to give out too many big loans. As a result, the weak force is felt only rarely, while the electromagnetic force is felt constantly all around us.

This distinction, however, holds only for particles without a great deal of energy. Imagine again the same two electrons, but this time they each have a million GeV of energy. With so much energy, they can each emit a Z boson with almost no effort—the 91.2 GeV needed to create such a particle is mere pocket change to these energy-rich electrons. In this case, exchanging a Z is no more difficult than exchanging a photon. Thus, for very energetic particles, the weak and electromagnetic forces are not much different. This would have been the case in the early moments of our universe's history, when all of space was incredibly dense and incredibly hot. All types of matter were present in the form of a thick soup of particles, virtually all of which possessed vast quantities of energy. In this early era, there was little difference between the electromagnetic and weak forces. They were truly fused together as a singular electroweak force. It is as if every investor had a billion dollars, just burning a hole in his or her pocket.

As the universe expanded and cooled, however, the energy required to generate the weak bosons became increasingly substantial. As energy became scarcer, the weak bosons were generated less frequently. With this transition, the weak force gradually became the feeble entity we see in our world today. The electroweak force was broken from its unified form in the first instants following the Big Bang into the two very different manifestations we see in our world today—the electromagnetic and weak forces.

IN ADDITION TO THE forces and the bosons intimately associated with them, the theory known as the Standard Model also contains a great variety of fermions. On the one hand, there are the quarks. In addition to the three quark species known to Gell-Mann—the up, down, and strange quarks—three other types have since been discovered: the charm, bottom, and top quarks.[3] These six particles can be found in each of the three colors, for a total of eighteen kinds of quarks and eighteen antiquarks. On the other hand, there are also the fermions known as leptons. The leptons include the electron and two other similar but heavier particles called the muon and the tau. Additionally, the Standard Model contains three leptons known as neutrinos. Neutrinos are the only known fermions that have neither electric charge nor color. Since the force of electromagnetism is only felt by particles with electric charge and the strong force only acts upon particles with color, neutrinos feel nothing besides the weak force and gravity. As a consequence, they almost never interact with anything. Although they are all around us, we rarely notice their ef-

---

[3] The bottom and top quarks are also sometimes called the "beauty" and "truth" quarks.

fects. Most of the time, neutrinos pass completely through Earth without ever interacting with it. Neutrinos are like ghosts in our world.

The fermions of the Standard Model can be divided up neatly into three sets, called families. In many ways, the three families are essentially copies of each other. The electron, the muon, and the tau, for example, play the same role in their respective families. Despite their very different masses—0.000511 GeV, 0.106 GeV, and 1.78 GeV, respectively—each of these three particles have the same amount of electric charge, spin, and other quantum properties. In this sense, the three families of fermions are like three models of the same thing, but built to different scales.

Why exactly multiple families are found in nature is one of the longest-standing open questions in physics. The first particle outside the first family to be discovered was the muon. When this heavy, electron-like particle first appeared in Carl Anderson's 1932 cosmic ray experiment, it was nothing short of perplexing (see chapter 3). No one had any reason to expect such a particle to exist. The positron, which was discovered only a few years earlier (also by Anderson), had a reason to be there—Dirac's equation predicted that it must. But in the 1930s, there were no such arguments to foresee the muon's existence. There are still no such arguments or explanations today.

Gradually, more and more second- and third-family fermions were discovered. In the mid-1970s, the tau lepton, the charm quark, and the bottom quark were discovered. Most recently, the last of the Standard Model fermions—the top quark—was discovered in 1995 at the Fermi National Accelerator Laboratory. This discovery completed the third family nicely. But although the fermions fit together neatly in the form of three families, no one knows why that number is three. Is there some reason it had to be three, rather than two, or even a single family? This is a mystery we have yet to solve.

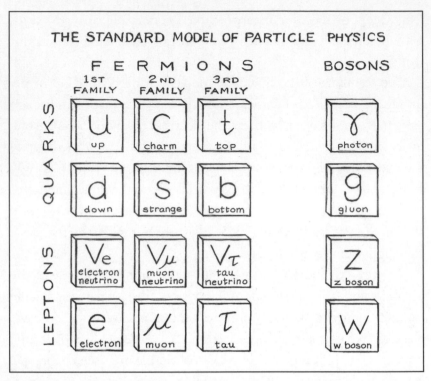

The fermions and bosons of the Standard Model.

There is also no known reason why the number of families has to stop at three. It is possible that some future particle collider experiment will discover particles from a fourth, or even higher, family. As it stands today, the existence of the three known families is simply an empirical fact without explanation. Perhaps someday we will understand the reasons for this fact, but for now it remains a mystery.

FROM WHAT I HAVE said so far about the Standard Model, it might appear to be a fairly ad hoc collection of particles and forces. In some ways, this is true. We don't have an explanation for why the symmetry of $SU(3) \times SU(2) \times U(1)$ is found in nature. We also don't

	COLOR	ELECTRIC CHARGE	MASS
Electron	No	−1	0.000511 GeV
Electron neutrino (v_e)	No	0	Less than about 0.0000000003 GeV
Muon (μ)	No	−1	0.106 GeV
Muon neutrino ($v_μ$)	No	0	Less than about 0.0000000003 GeV
Tau (τ)	No	−1	1.78 GeV
Tau neutrino ($v_τ$)	No	0	Less than about 0.0000000003 GeV
Down quark	Yes	−1/3	0.003 to 0.007 GeV
Up quark	Yes	+2/3	0.0015 to 0.003 GeV
Strange quark	Yes	−1/3	0.07 to 0.12 GeV
Charm quark	Yes	+2/3	1.25 GeV
Bottom quark	Yes	−1/3	4.2 GeV
Top quark	Yes	+2/3	171 GeV

Table 5. Some of the properties of the fermions of the Standard Model.

have an explanation for why the fermions of this symmetry come in three families. As far as we know, another symmetry group could have provided the underlying rules of our universe. Instead of three families of fermions, there could have been one, or two, four, or fourteen. For now, all we can say is that this is how we observe our world to be. In some ways, we simply don't know why it is the way it is.

It other respects, however, the nature of the Standard Model is not at all ad hoc. Basic requirements of self-consistency—through the principle of gauge symmetry—lead us to the conclusion that the Standard Model could not, in fact, have been very much different than it is.

When particle physicists calculate the probability that an event will take place—such as two particles interacting, or a particle decaying—they use helpful symbolic tools called Feynman diagrams. Feynman diagrams—named after their inventor, Richard Feynman—illustrate the various ways in which a given process can take place. For example, consider two electrons scattering or recoiling off one another through the force of electromagnetism. This can be represented by a simple Feynman diagram (shown right).

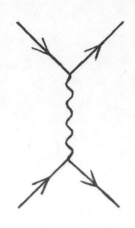

Here, the arrowed lines represent electrons and the wavy line a photon. The event that this picture describes is very simple. Two electrons come together, exchange a photon with one another, and then travel on. The only thing this process actually does is transfer some energy and momentum between the two electrons. In other words, the force of electromagnetism pushes or pulls between the two electrons.

But this is not the only Feynman diagram that plays a part in electron-electron scattering. On page 95 are some others.

In the upper-left diagram of that figure, for example, the same thing happens as I've just described, but as one of the electrons travels away it radiates another photon and then quickly reabsorbs it. In the upper-center diagram, a photon is radiated by an incoming electron and then is reabsorbed after another photon is exchanged with the other electron. In the lower-right diagram, the exchanged photon transforms into an electron-positron pair for a brief instant before converting back into a photon.

Each of these somewhat more complicated diagrams represents actual processes that can take place in the quantum world. Each and every one of these diagrams contributes to the calculation that determines how often electrons recoil off each other. And these are

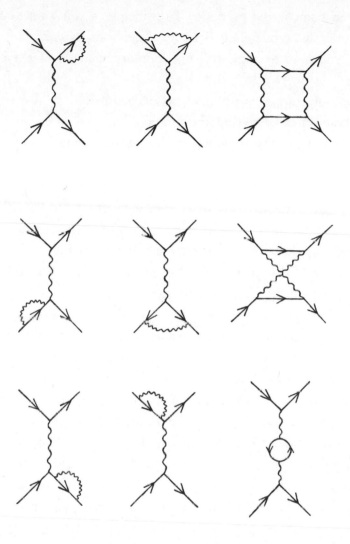

only the tip of the iceberg. The first diagram shown on page 94 is what is called a "tree-level" diagram. This captures the simplest way in which two electrons can interact. The other nine diagrams each contain a particle-loop somewhere in their structure. But there are also many other diagrams that I haven't written down with two loops, and even more with three, four, and five loops, all the way to infinitely complicated processes. In principle, if you want to calculate perfectly how likely it is for two electrons to scatter off each other, you have to sum the contributions from all of these diagrams.

This, of course, is no easy task. Fortunately, in the case of electron-electron scattering, the simplest diagrams are generally the most important. The diagrams with one loop typically contribute less than the tree-level diagram by a factor of a quantity known as α^2, which has been measured to be about $(1/137)^2$, or about 0.000053. The diagrams with two loops are even smaller—typically, about α^4 times less important than the tree-level contribution. Three-, four-, and five-loop diagrams produce effects that are smaller by factors of α^6, α^8, and α^{10}. In other words, the more complicated the diagram, the less likely that process is to occur.

So, to perform a given particle physics calculation, one needs to take the contribution from the tree-level diagram (or diagrams) and then add the smaller contributions from the several diagrams with one loop, plus the many even-smaller contributions from diagrams with two loops, plus the many, many very-much-smaller contributions from all of the three-loop diagrams, and so on. This wouldn't be so bad if it weren't for the fact that the more complicated, multi-loop diagrams are so numerous. On the one hand, each of the one-loop diagrams contributes much less than the tree-level diagrams do. On the other hand, there are many more loop diagrams than tree diagrams.

When all of this was first being worked out, it was not even clear whether these kinds of calculations would, at the end of the day, result in finite answers; it was feared that adding the contributions together from the infinite list of Feynman diagrams would yield a total of infinity. For these results to be finite is, of course, essential to the theory. The probability of two electrons scattering with each other—or for anything else to happen, for that matter—has to be finite. If adding up all of the possible processes resulted in an infinite probability, quantum field theory would be useless—not to mention nonsensical and wrong.

Fortunately, it was proven conclusively in the 1940s that all electromagnetic interactions yield finite probabilities when calculated in this way. This demonstration provided a much-needed vote of confidence in support of the theory of quantum electromagnetism. In order to perform this calculation, however, some rather strange steps had to be taken. When the contributions from the various individual Feynman diagrams for a given process are calculated, some of them—such as the upper-right and middle-right diagrams shown on page 95—yield perfectly finite and unproblematic results. Most of the others, however, do not. The contributions from these other diagrams go haywire and wildly diverge to infinity. The thing that Richard Feynman and others realized was that these divergent parts of the equations could be combined and thought of as part of the electron's electric charge. Through a mathematical process called "renormalization," the physics behind quantum electromagnetism can be shown always to be finite and sensible. Through this process, all of the divergent terms could be combined and reinterpreted in a perfectly self-consistent way. Although renormalization might seem like black magic—sweeping unwanted infinities under the rug—it is completely mathematically sound.

I guess we should not be surprised that a theory that so well describes an aspect of our universe is sensible. On the other hand, it is worth bearing in mind that many of the other theories proposed over the years have turned out to not be so trouble-free.

ALTHOUGH IT WASN'T CLEAR when the procedure of renormalization was first being formulated, particle physicists have since come to understand that this technique is intimately connected to the concept of gauge symmetry. Indeed, it is from this connection that gauge symmetries derive their staggering power. In order for the billowing sea of virtual particles—which exist throughout space and surround us everywhere—to make sense and to behave in a self-consistent way, very special characteristics must be present in the underlying theory. In particular, the theory must be able to be tamed through the process of renormalization—it must be renormalizable. Not all theories are. This simple requirement of self-consistency leads us back to the principle of gauge symmetry. We know now that theories respecting gauge symmetry are also renormalizable. In order for our world to make sense, it demands that it is built from these kinds of symmetries. It could have been no other way.

When Feynman and others were first attempting to discern whether the quantum force of electromagnetism was, in fact, self-consistent, they carried out their calculations assuming that the photon has precisely no mass. As far as we know, this is the case. No experiment has ever observed photons to possess any mass whatsoever. That being said, whether photons have zero mass or a very tiny mass remains, experimentally speaking, an open question. The photon, from an empirical perspective, could have an extremely small, but not quite zero, mass.

In any particle physics textbook, you can find the basic equation at the foundation of quantum electromagnetism—called the Lagrangian of the theory. Among other features, this equation describes a photon with exactly no mass. If we wanted to, however, we could simply add another term to that equation to change this, giving the photon any amount of mass we chose. But when we do this, we break the gauge symmetry of the theory. You might not object to gauge symmetry being broken in your theory, but I assure you that nature does. By giving the photon a mass, the divergent terms that used to nicely cancel each other so precisely no longer do. Suddenly, we have nonsensical and infinite predictions of our theory. With a massless photon, quantum electromagnetism makes perfect sense. But with even the tiniest photon mass, the theory's gauge symmetry is destroyed, and the whole picture falls apart completely and irreconcilably.

The moral of this story is that without gauge symmetry, there is no sensible way to understand the foundation of our world. This simple principle is built in and required of our universe's structure. It holds everything together and ensures its own self-consistency. To live in a world without it would be to live in a world where logic fails and no reality is possible.

GRADUALLY, THE SIGNIFICANCE AND power of gauge symmetry became increasingly understood and appreciated among physicists. By the middle of the 1960s, it had become the primary guiding principle used in particle physics and took an absolutely central role in the construction of the Standard Model. The strong force and eightfold way are a beautiful example of such a symmetry manifest in nature. The properties of the quarks and gluons are predicted by the patterns of the SU(3) gauge symmetry. The gluon, for example, is—much like the photon—required to be exactly massless by the principle of gauge symmetry. To the best of our experimental examination, this is precisely how these particles are found to be in our world.

In the early 1960s, a graduate student by the name of Sheldon Glashow began to think about the nature of the weak force within the context of gauge symmetries. His first theory proposed that the photon and the W^+ and W^- bosons were all part of the same larger symmetry group, and thus the electromagnetic and weak forces were manifestations of the same symmetry of nature. This first idea of Glashow's contained important mistakes, however, and ultimately turned out to be unworkable. For one thing, it was impossible to renormalize.

A year later, Glashow wrote about another theory designed to combine the weak and electromagnetic forces. He had discovered that if a fourth boson were included—the boson now known as the Z—along with the photon, the W^+, and the W^-, then the theory would be renormalizable.

At the time, there was no evidence for the existence of the Z boson. The charged weak bosons (the W^+ and the W^-) were seen indirectly through the manifestation of the weak force. For example, the process known as beta decay transfers electric charge, called a "charged-current" interaction. There were no analogous neutral-current processes known at the time. But despite the lack of experimental evidence, Glashow's hand had been forced to include this new phenomenon. The requirement of renormalizability insists that nature contain this extra particle. It is a beautiful example of the predictive power of gauge symmetry.

There was, however, a major problem with Glashow's theory of the weak interaction. Because of the observed range and strength of the weak force, the W and Z bosons each have to be very heavy particles. Glashow's original theory, however, is renormalizable only if the W and the Z are precisely massless—just as the photon and gluons have to be. To write down a simple theory with massive weak bosons means to write down a theory without gauge symmetry—and without gauge symmetry, nonsensical infinities are unavoidable. Glashow knew this problem was present, but he didn't know how to fix it.

This apparent contradiction made it difficult to build a self-consistent theory of the weak interactions. But difficulty is far from impossibility. The Standard Model contains a solution to this puzzle, in the form of a field permeating all of space—called the Higgs field—and a corresponding particle called the Higgs boson. If the Higgs field did not exist in our world, then the W and Z bosons would be massless—just as the photon and gluons are—and gauge symmetry would be restored. Through the presence of this field, however, the bosons of the weak force are given a great deal of mass while simultaneously respecting the gauge symmetry of the theory that is so clearly required.

MOST PEOPLE THINK OF mass as an intrinsic property of matter—something that cannot be given or taken away. Sure, people might put themselves on a bathroom scale each morning hoping to find that their own mass has been reduced—or at least not increased too much—but this kind of weight change is something that is only possible for composite objects and not fundamental particles. After all, an electron cannot lose or gain part of itself. An electron is a singular thing and doesn't consist of smaller parts to gain or lose. For this reason, a particle's mass seems like it should be an immutable property. This, however, is not always the case. To understand why, reconsider for a moment what mass actually is, and what causes it to exist.

According to Albert Einstein and his theory of relativity, energy comes in two basic forms. First, it can take the form of motion—called kinetic energy. But second, it can also come in the form of stationary energy, or what we call mass. Einstein's equation $E=mc^2$ describes the relationship between mass and energy, and shows us that mass can even be created from energy and vice versa. Two photons—which themselves have no mass—can produce a pair of massive particles— an electron and a positron, for example—if they are collided together with enough energy. The reason that physicists build accelerators that collide particles together is to put enough energy together in one place at one time to create new forms of matter. The more massive those forms of matter are, the more energy is needed in the collision. It all comes back to $E=mc^2$.

In some cases, the distinction between kinetic energy and mass energy is not so clear. Take, for example, the quarks bound together to make up a proton. Protons have been measured to have a mass of about 0.938 GeV. The two up quarks and the one down quark that make up a proton, however, are much, much lighter—a few thousandths of a GeV each. If I bind three one-pound stones together, I would expect the total object to weigh an amount equal to the sum of its parts—three pounds. Strangely, the masses of the constituent quarks do not come close to accounting for the total mass of a proton. It is as if the three one-pound stones together weighed over a thousand pounds.

The trick to resolving this conundrum is to remember that mass is nothing more than a form of energy, and energy can be created by motion. The quarks inside of protons are moving very quickly, and possess a great deal of kinetic energy. So, in addition to the quark masses themselves, the internal motion of the quarks and gluons is where the vast majority of the proton's mass comes from.

A somewhat similar process can provide the W and Z bosons with mass, without breaking the gauge symmetry of the theory. But to do this, there has to be another source of kinetic energy. That kinetic energy comes from interactions with a new particle, called the Higgs boson. As a W or a Z boson travels, it constantly feels the influence of the Higgs field that permeates all of space. Higgs bosons are brought into existence everywhere constantly. If somehow we could turn off the Higgs field, every W and Z boson would instantly become massless—and the weak force would become long ranged and about as strong as the force of electromagnetism. In the presence of this field, however, the W and Z particles become burdened and heavy. It is even thought that each of the fermions of the Standard Model—electrons, muons, quarks, and so on—acquire their masses through their interactions with the Higgs. In many ways, it is this mysterious particle that brings our universe into the form in which we find it. It is for this reason that the Higgs boson has earned its nickname, the "God Particle."

The Higgs boson is the only particle within the Standard Model that has never been observed in experiments. This is not for lack of trying. Experiment after experiment has been performed in search of this elusive particle, each one ruling out Higgs bosons with another range of masses. The theory behind the Standard Model does not predict exactly how heavy the Higgs boson should be. In order to be self-consistent, its mass should be less than about a thousand GeV, but it could also be much lighter. In fact, indirect evidence even suggests that it should be much lighter—less than about 144 GeV. The experiment that has searched for the Higgs boson over the highest range of masses is LEP—the Large Electron-Positron Collider. Data from LEP has ruled out the possibility of the Higgs boson being lighter than 114 GeV. The region in which the Higgs is thought to be hiding is slowly shrinking. It has remained hidden from us for a long time, but that time is coming to an end.

In the fall of 2000, the LEP experiment was shut down to make room for the Large Hadron Collider, which is being built on the same site. But even in the absence of LEP, the search for the Higgs boson goes on. At the time I first wrote this paragraph, it was the summer of 2007 and rumors had been circulating throughout the physics community about a signal seen at the Fermilab Tevatron accelerator—a signal that looks like a Higgs boson. These kinds of rumors appear from time to time, sometimes being the first indication of a true discovery, but oftentimes not. Such a signal often begins with just a handful of strange and unexpected events that could either be a detection of something new or a mere statistical fluctuation resulting from ordinary processes. As time goes on and more data is collected, however, these questions eventually become resolved and firm conclusions reached. By the time this book is published and you are reading it, either a new particle will have been discovered at Fermilab with great fanfare or these rumors will have calmed down and gradually died away. I'm crossing my fingers for the former.

Where is the idea of a Superworld
coming from? Could half of the
particles, at least, have escaped our
direct observations?

—*Pierre Fayet*

|||||||||| **6** ||||||||||

The Birth of Supersymmetry

In the fall of 2000—during my second year as a graduate student at
the University of Wisconsin—I attended my first particle physics
conference. The meeting, titled "SUSY 30" in reference to the thirty
years that had passed since the first work on supersymmetry, was
held in Minneapolis, on the University of Minnesota campus. To
me, SUSY 30 looked like the ideal meeting to attend. For one thing,
over the previous several months I had been working on a research
project with my adviser and another physicist on the topic of supersymmetry. Attending a conference on that subject would almost
certainly help me to understand aspects of my project that were still
confusing to me. Second, the trip from Madison to Minneapolis was
a quick four-hour drive, and I knew a friend who lived near campus
who would probably let me sleep on his couch.[1] For a second-year

---

[1] Thanks, Tom!

graduate student, getting money from your adviser or university department to travel to conferences can sometimes be a tricky thing to accomplish. The Minneapolis meeting offered an opportunity for me to learn more about the current research on supersymmetry—and on the cheap.

The Minneapolis meeting turned out to be a great experience for me. The first few days were filled from morning until night with presentations on various aspects of supersymmetry and related physics. Then, for about the two weeks that followed, the meeting transformed into a workshop atmosphere where people worked and discussed physics without too many seminars to attend—typically, one or two hour-long talks each day. For me, these relaxed days provided a great chance to talk with many far-more-experienced particle physicists. I learned a great deal over those two weeks.

When I started to write this book, I decided to read over the proceedings from the Minneapolis meeting. Just as I had remembered, they are full of papers describing aspects of the current—as of 2000—research on supersymmetry. In addition to these, however, there were also a number of other contributions to the conference's proceedings. To my surprise, these papers recalled and retold the history of the supersymmetric theory. Many, if not most, of the key figures in the development of supersymmetry had come to Minneapolis for this meeting to reminisce about the beginnings of the theory. From the Soviet side of supersymmetry's origin, there were talks by Evgeny Likhtman, Vladimir Akulov, and Vyacheslav Soroka. Although some of the Soviet founders of supersymmetry had passed away since its birth, their contributions were recounted by others: Natasha Koretz-Golfand, the wife of Yuri Golfand, discussed the life and work of her late husband, and Vladimir Akulov discussed Dmitry Volkov's early work on supersymmetry.

From the other side of the Cold War divide, talks were given by many of the leading figures in supersymmetry's development, including Pierre Raymond, John Iliopoulos, John Schwarz, Gabriele

Veneziano, Pierre Fayet, and many others.[2] As an inexperienced student, I had no idea that I was surrounded not only by scientists actively researching supersymmetry, but also by many of the scientists who had participated in inventing it.

UPON GOING BACK OVER the proceedings of the Minneapolis meeting, one of the contributions I enjoyed reading the most was written by the French theoretical physicist Pierre Fayet. Fayet's paper wasn't about the origins of the theory of supersymmetry but instead focused on the period of research that came immediately afterward. Supersymmetry's original conception—in both the East and the West—took a very abstact and mathematical form. In the 1970s, it was not at all clear that supersymmetry could really have anything to do with the real world—even to the physicists working on it. Although supersymmetry was certainly a natural and attractive extension of the mathematical symmetries found in quantum field theory, it was very far from clear that it could be formulated in a way that could actually be present in our universe.

Fayet's talk at the Minneapolis meeting described the period in the late 1970s when physicists began to take the presence of supersymmetry in our world far more seriously. Over these years—from around 1974 to 1981—this theory was transformed from an esoteric mathematical concept into a concrete theory describing particles and interactions that could potentially be observed. During this time, supersymmetry took a very large step toward reality.

IF SUPERSYMMETRY EXISTED AS a perfect symmetry of nature, then our world would be a very different place. In our world, we see

---

[2] As far as I know, the two foremost pioneers of supersymmetry in the west—Bruno Zumino and Julius Wess—did not attend the meeting.

electrons—every atom contains them—but we see no boson counterpart to the electron. Every time we open our eyes we see photons,
but no photonlike fermions. If you think of supersymmetry as a
mirror, with bosons and fermions as reflections of each other, it is
strange that our world seems to contain some particles on each side
of the mirror—some fermions and some bosons—but none of those
particles' reflections. The superpartners seem to be missing from
our world.

Dirac's symmetry between matter and antimatter has no such
missing reflection. As far as we know, Dirac's symmetry is an absolute symmetry in this regard. For every kind of particle we see in our
world, there is a corresponding antiparticle, with precisely the same
mass. Experiments have confirmed this reality over and over again.
Dirac's symmetry between matter and antimatter is found perfectly
manifest in nature.

Unlike Dirac's symmetry between matter and antimatter, however, supersymmetry between bosons and fermions must be a broken symmetry. If supersymmetry in fact exists in nature, then the
new superpartner particles that are predicted by this theory must be
considerably heavier than the particles we know and observe in our
world. The superpartner particle of the electron—called the super-
electron, or selectron—has to be at least two hundred thousand
times heavier than the electron, for example. If it were not, accelerator experiments would have created and discovered these particles
by now. So far, they have not.

This may strike you as a blemish on the elegance and appeal of
supersymmetry. Many physicists feel the same way. A perfect and
unbroken symmetry is a truly beautiful thing to behold. That being
said, perfect symmetries are very rare in our world. We find examples of imperfect, or broken, symmetries almost everywhere we look
in nature. The world is full of symmetries, the vast majority of which
are broken to some extent. Supersymmetry might not exist as a per-

fect and unbroken symmetry of our world, but it may very well take the form of one of our universe's many broken symmetries.

Consider for a moment an example of symmetry that you experience in the world everyday: the human face. Faces are approximately left-right symmetric—if you draw a line down the center of your face, each side looks like an approximate mirror reflection of the other. This symmetry is not at all perfect, however. Slight differences between the left and right sides of a person's face break the symmetry. In fact, if you were to take an image of one side of your face and place it alongside its mirror reflection, the resulting image will appear strange and unnatural. We somehow instinctively expect the symmetry of a human face to be broken. Perfect symmetry in biology can sometimes appear strange and otherworldly, and even leave us feeling uncomfortable or suspicious.

A human face possesses an approximate, but imperfect, left-right symmetry. In the right frame, the same face is shown, but with its right side replaced with a mirror reflection of the left side—in other words, the perfect left-right symmetry has been restored. The appearance of this strikes most people as strange and unnatural.
Photo credit Hebrew University of Jerusalem.

The shape of our planet Earth is one of my favorite examples of a broken symmetry. The force of gravity has formed Earth into an almost, but not quite, perfect sphere. If you could hold Earth in your hands, it would feel as perfectly round as a new bowling ball. The tallest mountains on Earth are less than 30,000 feet above sea level. This is absolutely tiny in comparison to the planet's radius of 20,853,000 feet. Even the tallest mountain peak is only about one-tenth of 1 percent farther from the center of Earth than sea level is. This is an absolutely tiny departure from our planet being a perfectly smooth sphere. Earth's hills, valleys, and mountains break its spherical symmetry, although only slightly.

Philosophers have long held geometrical symmetry in high regard. Aristotle saw Earth as an imperfect realm, full of broken symmetries. In contrast, he maintained that the heavens—meaning the moon, the planets, and the stars—must consist of only the most perfect forms, which he argued must exhibit absolute and perfect symmetry. This viewpoint was adopted and maintained by many, and for many centuries was held as the official doctrine of the Roman Catholic Church. This turned out to be an awkward and difficult position to stand behind when a relatively obscure mathematician by the name of Galileo Galilei published a short volume titled *Sidereus Nuncius*—or *Starry Messenger*—in 1610. Galileo's book described the first astronomical observations performed with a telescope, including the discovery of four of Jupiter's moons and mountains upon the surface of the Moon. The heavens, it was revealed, possessed the same kinds of broken symmetry that were found on Earth. Galileo's observations also supported the Copernican system in which Earth revolved around the Sun, and not vice versa. Sadly, the Church stood behind its ancient and flawed doctrine, even in the light of Galileo's observations to the contrary. For his discoveries, Galileo was eventually tried by the inquisition for heresy, silenced, and confined to house arrest for the remainder of his life.

The leaders of the Church didn't want to believe that the heavens are imperfect, but they are nevertheless. Whether we find it appealing or not, many of the symmetries around us are in fact broken. Supersymmetry must be broken if it is to exist in our world.

I HAVE SAID BEFORE in this book that supersymmetry predicts that many new varieties of particles must exist. But what are these new forms of matter like? What properties do they have? What kind of experiments might we be able to build to search for them?

In late 1981, physicists constructed for the first time a realistic theory that included the superpartner particles predicted by supersymmetry. This first such supersymmetric model, proposed by Howard Georgi at Harvard and Savas Dimopoulos at Stanford, is known as the Minimal Supersymmetric Standard Model, or the MSSM for short. The MSSM is the most simple form that supersymmetry could possibly take in our world.

Within the context of the MSSM, the characteristics of the supersymmetric partners of the Standard Model particles are fairly straightforward to deduce. For each Standard Model particle, there are superpartners with all of the same quantum properties—electric charge, color, and so on. Exactly how these particles are accounted for depends on a property known as spin.

The spin of a particle is a quantity that has to do with how much intrinsic angular momentum it can possess. An object that is rotating around its own axis—like Earth or a top—has a certain amount of angular momentum. Even though fundamental quantum particles are pointlike and, therefore, don't have a center to rotate around, they, too, can possess their own angular momentum. This is just another bit of weirdness in the quantum world.

Spin is also an essential element behind the theory of supersymmetry. Recall that spin is, in fact, what determines whether a particle is a fermion or a boson. Particles that have half-integer spins—such as 1/2 or 3/2—are fermions. Those with integer spins—such as 0 or

1—are bosons. Particles with a given spin can possess only specific amounts of angular momentum. A fermion with a spin of 1/2 has two possible angular momentum configurations, called the +1/2 and –1/2 states. For a boson with a spin of 1, three states are allowed if it is massive (+1, 0, and –1) but only two (+1 or –1) if it has no mass. A particle with 3/2 spin allows for four different states (+3/2, +1/2, –1/2, and –3/2) and so on.

According to the mathematics of supersymmetry, the spins of a particle and its superpartner must differ by 1/2. The electron's superpartner, for example, must have a spin of either 0 or 1. To determine which of these is the case, we have to match up the number of possible boson and fermion states. Electrons have two such states (+1/2 and –1/2). If the selectron had a spin of 1, then it would have three (+1, 0, and –1). Two and three do not match, thus excluding the possibility of a spin-1 selectron. If we consider the other possibility, in which the selectron had a spin of 0, we have only one selectron state. Therefore, to match this up with the two states of the electron, we have to introduce not one, but two, spin-0 selectrons.

We can perform this same routine with the entire Standard Model and, for most of the particles, it works out in much the same way. There are a couple of complications, however. Neutrinos, for example, only exist in half the forms that the other fermions do, so instead of two, only one variety of sneutrino is needed (for each of the three families).

Another peculiarity appears when considering the superpartners of the photon, Higgs bosons, and gauge bosons.[3] In Table 6, the superpartners of the photon, the Z boson, and two of the Higgs bosons—also known as the photino, the zino, and Higgsinos—are each referred to by the same name: "neutralino." Similarly, the wino

---

[3] You might notice here and in Table 6 that there is not only a Higgs boson, but several kinds of Higgs bosons—called the h, H, A, H$^+$ and H$^-$. For technical reasons which I won't go into here, this turns out to be a requirement of supersymmetry.

PARTICLE	COLOR	ELECTRIC CHARGE	SPIN	SUPERPARTNER(S)	SPIN
Electron	No	−1	1/2	2 Selectrons	0
Electron neutrino (v_e)	No	0	1/2	1 Electron Sneutrino	0
Muon (μ)	No	−1	1/2	2 Smuons	0
Muon neutrino (v_μ)	No	0	1/2	1 Muon Sneutrino	0
Tau (τ)	No	−1	1/2	2 Staus	0
Tau neutrino (v_τ)	No	0	1/2	1 Tau Sneutrino	0
Down quark	Yes	−1/3	1/2	2 Down Squarks	0
Up quark	Yes	+2/3	1/2	2 Up Squarks	0
Strange quark	Yes	−1/3	1/2	2 Strange Squarks	0
Charm quark	Yes	+2/3	1/2	2 Charm Squarks	0
Bottom quark	Yes	−1/3	1/2	2 Bottom Squarks, or Sbottoms	0
Top quark	Yes	+2/3	1/2	2 Top Squarks, or Stops	0
Photon, Z, and the Neutral Higgs Bosons	No	0	1	4 Neutralinos (Photino, Zino and 2 Higgsinos)	1/2
Gluon	Color Anticolor	0	1	1 Gluino	1/2
W+/− and Charged Higgs Boson (H+/−)	No	+1/−1	1	2 Charginos (Wino and Charged Higgsino)	1/2

Table 6. The particles of the Minimal Supersymmetric Standard Model.

and the charged Higgsino are both called "charginos." The reason for this strange multinaming convention is that each of the four neutralinos and two charginos is not expected to be the superpartner of uniquely one kind of particle. Each of the four neutralinos is instead a combination of the photino, the zino, and the two Higgsinos. Together, these four neutralinos make up the superpartners of the photon, Z, h, and H, but only collectively. In a similar fashion, the two charginos are together the superpartners of the $W^{+/-}$ and the $H^{+/-}$, but individually are mixtures of the two.

I have a favorite analogy for explaining the messy mixture of states that are the neutralinos and charginos. Neutrinos and sneutrinos are like a conventionally married couple—one neutrino partnered with one sneutrino. Electrons and selectrons have, in contrast, something of a polygamous arrangement—for the electron, there are two selectron partners. The same is true for muons, taus, quarks, and their superpartners. The four neutralinos, instead of being connected to individual particles, are collectively partnered to four ordinary particles, like a group marriage of four men and four women. The $W^{+/-}$, the $H^{+/-}$, and the two charginos are in a similar arrangement. It would seem that there is not much adherence to traditional family values in the supersymmetric quantum world.

WITH THE DEVELOPMENT OF a realistic supersymmetric model— the MSSM—it had finally been demonstrated that supersymmetry was not only an elegant extension of nature's blueprint, but also a plausible one. But it was not for beauty or even plausibility that supersymmetry became the immensely popular theory that it is today. Among other reasons, many physicists came to learn about and study supersymmetry because, despite all of its successes, the Standard Model suffers from a sickness. The cure for this ailment, it turns out, may be supersymmetry.

Although it has never been shown to be incorrect in any of its predictions, the Standard Model as we understand it is ultimately unstable and is in desperate need of a new mechanism to prevent it from falling apart. This perplexing problem can be traced back to the one particle of the Standard Model that has still never been detected in any experiment—the elusive Higgs boson.

In the last chapter, I described how the particles of the Standard Model have their masses generated through their interactions with the Higgs boson. If it were not for the presence of the Higgs field, it is thought that the Z and W bosons, electrons, muons, top quarks, and other particles would each be entirely massless. But this mechanism is also capable of working in the opposite direction. The particles of the Standard Model return the favor to the Higgs, generating a mass for it as well. When the contributions from all of the known particles are accounted for, it seems that the Higgs boson should be very, very heavy—much heavier than it could possibly be.

Roughly speaking, each kind of particle should give a truly colossal contribution to the mass of the Higgs—roughly in the ballpark of 10^{15} GeV or more in many cases. Although the exact size of this number depends on unknown details of the theory, the point is that anything remotely like 10^{15} GeV is dramatically larger than the Higgs boson appears to be—according to the current set of indirect measurements, it can be no heavier than a couple hundred GeV or so. The only way that the Higgs could possibly be this light is if the contributions from each of the various kinds of particles somehow cancel each other out almost exactly. This is not completely impossible because bosons and fermions each contribute to the Higgs mass with opposite signs—they work against each other in making the Higgs heavy. Therefore, if the boson contributions to the Higgs mass are almost exactly the same size as those coming from the fermions, this sickness could potentially be cured.

This doesn't work very well in the Standard Model, however. In this situation, there is essentially a collection of positive and negative contributions to the Higgs mass—many of which are as large as roughly 10^{15} GeV or so. When you add these various contributions together, it is wildly unlikely that they would all cancel out to something as small or smaller than 144 GeV. To give you an idea of how implausible such a coincidental cancellation would be, imagine rolling a trillion dice and adding up all of the numbers. You should expect to get a number around 3.5 trillion. Now imagine rolling the trillion dice again, and subtracting the new result from the first. In all likelihood, you will find a fairly large difference—something in the millions. In order for the two dice rolls to match each other as precisely as the fermions and the bosons would have to match each other to obtain an acceptable value of the Higgs mass, you would have to roll almost identical numbers, like 3,500,002,536,572 and 3,500,002,536,571. This is not impossible, but it is exceedingly unlikely.

This problem with the Higgs boson's mass is often referred to as the "hierarchy problem." There is an enormous gap—or hierarchy—between the size of the Higgs mass—no more than a couple hundred GeV—and the size of the various contributions to this mass, which are 10^{15} GeV or so. Either some mechanism must come into play to prevent the Higgs from becoming exceedingly heavy, or the mathematical details of our universe's structure must be tweaked in such a way that—against all the odds—all of the contributions to the Higgs mass almost perfectly cancel each other. One of the biggest challenges facing particle physics today is to find out how nature manages to accomplish this. What makes the Higgs so light? Is there a reason for our world to be this way, or is it just dumb luck?

Supersymmetry to the rescue.

THE PRESENCE OF SUPERSYMMETRY in our world provides a simple solution to the perplexing hierarchy problem. I said before that the contributions to the Higgs mass from fermions and bosons work against each other. Furthermore, if a boson and a fermion have the same quantum properties, their contributions to the Higgs mass will also be of the same size.

Take, for example, the part of the Higgs mass acquired through its interactions with the top quark. Fermions such as the top quark contribute through diagrams like this one.

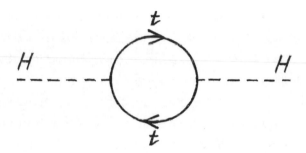

Here, the Higgs boson is spontaneously converted into a top quark and an anti-top quark, although only for an instant. Processes such as this one occur constantly for all Higgs bosons, and sustain its mass.

If the world is supersymmetric, we can also write down another contribution from the top squarks.

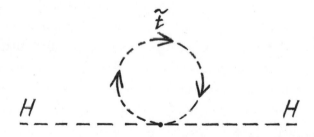

If the top quark and the top squarks have the same electric charge, color, and other quantum properties—just as supersymmetry predicts them to have—then these two sets of diagrams will almost perfectly cancel each other's effects. In fact, if the masses of the top quark and top squark are the same—if supersymmetry is unbroken—then the quark and squark contributions cancel exactly. But even if supersymmetry is broken, this cancellation can be sufficiently good to control the Higgs mass—as long as the superpartner particles are not too heavy.

Diagrams similar to the two on page 117 can be written down for every kind of particle and its superpartner. Every kind of fermion-antifermion pair can be created and destroyed just as in the first diagram, and every kind of boson adds a contribution like that shown in the second.

Through this connection between bosons and fermions, supersymmetry manages to fix the results of the dice roll in our earlier analogy. When the trillion dice are rolled for the first time, this corresponds to the contribution to the Higgs mass from diagrams with a fermion loop, like the one with top quarks shown on page 117. If not for supersymmetry, we would roll the trillion dice once again to determine the size of the contribution from diagrams with a boson loop. But with supersymmetry, we don't have to roll a second time at all. The dice are fixed. The second roll *has to be* almost identical to the first. The residual left behind after the two rolls are subtracted from one another is, as a consequence of supersymmetry, tiny, and, therefore, the mass of the Higgs boson is small—just as it needs to be.

In this way, the principle of supersymmetry ensures that the boson and fermion interactions with the Higgs boson leave our world in a form consistent with what we observe. Supersymmetry provides the best solution to the hierarchy problem we have ever conceived of. On its own, the Standard Model is very sick. When combined with supersymmetry, it is healthy and strong.

———

THERE IS AN AMAZING degree of flexibility in the quantum world. Almost anything you can imagine happening does, on occasion. Particles can come forth out of nothingness. Forms of matter can even transform into others. The very nature of energy, space, and time make almost everything possible in the quantum world.

But not quite every process actually takes place in our world. We have never witnessed the creation or destruction of electric charge, for example. When an electron is created, it is created alongside a positron, for a net change of zero electric charge. As far as we know, electric charge is always conserved—period.

Nature's strict adherence to the law of conservation of electric charge also explains why electrons, unlike most other kinds of particle, never decay. Electrons are the lightest of all electrically charged particles. If an electron were to decay into other forms of matter, it would either have to leave behind heavier types of particles, which would violate the law of conservation of energy, or lighter but electrically neutral particles, which would violate the law of conservation of electric charge. Electrons are thus, to the best of our knowledge, infinitely stable.

From humanity's standpoint, it is very fortunate that electrons are stable. The vast majority of the electrons in our universe have existed since the first instants following the Big Bang. If they had decayed during the billions of years since, we would have almost no electrons in our world today. And if we were without electrons, we would also be without atoms. It is hard to imagine life existing in such a universe.

But electrons are not the only stable ingredients that are needed to build atoms—we need protons as well.[4] Unlike in the case of

---

[4] Neutrons are also found in atoms, but are not themselves stable. They are only stable inside atomic nuclei. When left in isolation, neutrons decay within a matter of minutes into a proton, an electron, and an antineutrino.

electrons, however, proton stability is not guaranteed by the conservation of electric charge. We can imagine processes that could cause a proton to decay without changing the total electric charge—a proton transforming into a positron and photons, for example. But, for reasons unknown, this hypothetical process has never been observed. To the best of our knowledge, protons never—or at least almost never—decay.

In the Standard Model, proton stability is automatically guaranteed. There are no processes in the Standard Model that change the total number of baryons present.[5] In much the same way that electric charge is a conserved quantity, baryon number cannot change within the context of the Standard Model. Quarks can be created, but only along with antiquarks. A proton cannot be destroyed without also destroying an antiproton or some other kind of antibaryon. According to the Standard Model, the proton can expect to live an incredibly long life.

Theories building upon and extending the Standard Model, however, offer no such guarantees. For example, according to theories that attempt to unify the three forces of the Standard Model—called grand unified theories—protons are predicted to decay, although only very, very slowly. In the simplest grand unified theories, protons are generally predicted to exist for around 10^{30} years before they decay. This is a mind-bogglingly long time. Considering that only about 10^{10} years have passed since the Big Bang, we should not be surprised that the overwhelming majority of protons have survived to the present day.

Proton decay is not, however, a prediction limited to grand unified theories. Although supersymmetry can provide a cure for the

---

[5] This is an almost true statement. There is, however, the possibility of extremely rare processes leading to proton decay according to the Standard Model, but at a rate far too slow to be observable.

sickness of the hierarchy problem, there is also a danger of side effects. In particular, according to the first incarnations of supersymmetry, protons should decay—and quickly. Since we don't see this happening in our world, this prediction does not bode well for any theory that makes it. In the case of supersymmetry, the basic problem is that new kinds of processes involving superpartner particles could potentially cause protons to decay. For example, the existence of squarks could cause protons to decay into a positron and a pion.

Such a process—if it were allowed to take place—would likely cause most of the protons in our universe to decay within a fraction of a second. If care were taken to construct a supersymmetric model that suppressed this process as much as possible, the proton's lifetime could be pushed up to about a year, but not much longer. Such a universe would be violent beyond imagination, and very far from habitable. If squarks and other superpartner particles exist in our universe, then something has to constrain their interactions so that this kind of process cannot take place.

Fortunately, there is a simple way to do exactly this. In particular, the prevention of proton decay in supersymmetric models can be ensured by a physical law, known as the conservation of R-parity. R-parity is simply a value, or number, that every kind of particle has—just like electric charge, spin, strangeness, and many other quantum properties. The R-parity of each kind of particle is determined by a simple formula: $P_R = (-1)^{3(B-L)+2S}$. Here, the symbols B, L, and S denote the particle's baryon number (+1/3 for all quarks, −1/3 for antiquarks, and 0 for all others), lepton number (+1 for all leptons, −1 for antileptons, and 0 for all others), and spin (1/2 for fermions and 1 or 0 for bosons). When we calculate the R-parity of *any* particle in the Standard Model, we always find a value of +1. The R-parity of any given superpartner is, in contrast, always −1.

The basic rule behind this conservation law is that in any interaction or process, the product of all of the particles' R-parity values must never change. Within the Standard Model, this rule changes absolutely nothing. All of the particles of the Standard Model have R-parity of +1, so when you multiply different combinations of them together, you always get +1. Thus, every Standard Model process imaginable automatically respects the law of conservation of R-parity.

The same is not true, however, once we throw the superpartners into the mix. Take the squark-exchanging process, shown in the previous figure, that leads to proton decay, for example. At the start, there are three quarks present, for a total R-parity of $+1 \times +1 \times +1 = +1$. When two of the quarks are transformed into a squark, the R-parity becomes transformed to $+1 \times -1 = -1$, thus changing the total R-parity. If nature is forced to follow the law of R-parity conservation, then this process cannot possibly take place. In fact, it follows that superpartners can only be created or destroyed in pairs, and never in isolation. Without R-parity conservation, supersymmetry says our world should have disintegrated long ago. With this additional law, however, the universe is safe and quite stable.

Since the introduction of R-parity in the 1970s, it has been found to be of great importance in many applications beyond the mere prevention of proton decay. As the measurements of the Standard Model particles and their interactions have become increasingly precise, we would have expected to see the effects of superpartner particles if they were not constrained by the rules of R-parity conservation. If R-parity is conserved, however, the processes that these measurements are the most sensitive to are conveniently forbidden. The fact that we still have not found any compelling evidence for any non–Standard Model processes is—in addition to proton stability—another strong reason for us to think that R-parity is conserved in our universe.

Yet another reason for us to believe that R-parity is conserved comes from our universe itself. Astronomers have found that the vast majority of matter in our universe is an invisible substance—or substances—called dark matter. If R-parity is conserved, then the lightest of the superpartner particles cannot decay, and could very likely constitute this mysterious substance. Without R-parity conservation, we would have to look beyond supersymmetric theories for a solution to this cosmological conundrum.

R-parity, introduced three decades ago to save the proton, seems today to be needed to explain supersymmetry's elusiveness, and to explain why so much of our universe's matter is elusive. In physics, good stones always seem to kill lots of birds.

IN 1981, A COMPLETE and self consistent theory of supersymmetry was finally in hand. When Howard Georgi and Savas Dimopoulos first proposed their Minimal Supersymmetric Standard Model—the MSSM—it included all of the aspects of supersymmetry that I have written about in this chapter. They knew they had in their hands a solution to the instability of the Higgs mass—the hierarchy problem. They also knew that R-parity conservation must be included if their model were to be a realistic one.

Georgi and Dimopoulos's model said a lot about how they imagined supersymmetry to be found in our world. But it also left a great deal unsaid. We don't know precisely how heavy the superpartner particles are, or exactly how they will interact with each other. All told, the MSSM has a whopping 120 free parameters—numbers with unknown values, such as the masses of the various superpartner particles and other properties of the theory. Although a great deal of progress had been made in understanding how supersymmetry might be manifest in our world, it could still take a great many forms.

In the almost thirty years since the MSSM was first introduced, little progress has been made in our understanding of how supersymmetry is expected to appear. Although several other more complicated supersymmetric models have been written down and studied, and a vast number of calculations have been performed to predict how supersymmetric particles will behave and what effects they might have, these efforts have not dramatically changed how we think about supersymmetry. The theoretical physicists studying supersymmetry are at something of an impasse.

I sincerely doubt that additional theoretical speculation will contribute substantially more to answering these questions. To breach this impasse, we will need empirical input. Theorists have come up with numerous ways in which supersymmetry might appear in our world. It is up to experimenters to tell us which of these, if any, is right. After all, we are talking about science and not theology or philosophy. In science, observation resolves all debates.

**All things come out of the one and the
one out of all things.**
—Heraclitus

**It is a wonderful feeling to recognize the unifying
features of a complex of phenomena which present
themselves as quite unconnected to the direct
experience of the senses.**
—Albert Einstein

|||||||||| **7** ||||||||||

Unity in All Things

I would argue that the Standard Model of particle physics is the single most successful theory in the history of science. It is a theory that makes numerous precise predictions regarding the characteristics and behavior of the matter and force found in our world. It can be used to predict how long it takes for a species of particles to decay, or the probability that different particles will interact with each other. In some cases—such as the top quark—it has even been used to reveal that a new species of particle must exist, well before it was first observed or discovered. With the lone exception of the force of gravity, the Standard Model explains the foundations of all phenomena known to exist in our universe.[1] In the decades since its

---

[1] When it comes to understanding the force of gravity, physicists rely on the theory of general relativity. It is known, however, that this theory is ultimately incompatible with the laws of quantum physics. Efforts to develop a viable quantum theory of gravity have, to put it mildly, been unsuccessful thus far.

conception, not a single one of the Standard Model's many predictions has ever been shown to be incorrect. By any measure, it has been an astonishingly successful theory.

But despite the many successes of the Standard Model, there are questions that this theory leaves unanswered. For example, the problem of how the Higgs boson is prevented from becoming ultra-heavy, as discussed in the previous chapter, is left unresolved in the Standard Model. If this catastrophe is to be averted, the Standard Model must be extended in some way—such as through supersymmetry. But there is another kind of question that the Standard Model leaves unaddressed as well. In particular, where does the structure of the theory itself come from? Why does it take the form we find it in? Why does it contain three forces, and not one, two, or fourteen? Why are there three families of fermions? Why do the patterns and structures we observe in our universe exist?

The Standard Model describes the underlying structure of our world as a manifestation of a fairly complex set of symmetries—what particle physicists call $SU(3) \times SU(2) \times U(1)$. There is little doubt that this pattern describes our world very well—experiments confirm it again and again. But despite the confidence we have in the Standard Model's predictive power, we would also like to understand why these symmetries are present in our world, and where they come from. As the experiments of the 1970s and 1980s confirmed the validity of the Standard Model in ever-greater detail, more physicists began to ask questions of this nature. As the Nobel Prize–winning physicist Frank Wilczek remarked, "It was no longer a question of testing $SU(3) \times SU(2) \times U(1)$, but of explaining it." The focus of many particle physicists shifted from understanding the Standard Model itself to understanding where it came from. In a sense, they sought to understand the ancestry of the Standard Model.

THE SYMMETRIES OF THE Standard Model collectively bring forth each of the three forces we experience in our world.[2] The SU(3) part of the Standard Model describes the nature of the strong force. The SU(2) and U(1) symmetries each contribute to the electromagnetic and weak forces. How these symmetries came to be forged into our universe's structure has been the central question in the careers of many physicists. These physicists envision a single overarching symmetry from which each of the forces of the Standard Model emerges. These theories are based upon an idea known as grand unification. The efforts to build a grand unified theory—a GUT—that successfully describes our universe have been among the most monumental undertakings of modern science.

But how can the multiple symmetries of the Standard Model come out of a single grand unified symmetry? What does it mean for one kind of symmetry to "contain" others? To get a better idea of what I mean by this, imagine a simple geometrical example. Picture a perfectly flat and round disk. A perfect disk is the same, regardless of which side is facing up. We can tell which side of a quarter is facing upward because it has a heads or tails engraved on it, while the two sides of a perfectly symmetric disk are identical. The indistinguishability of the two sides of a perfect disk is one of its symmetries. But the perfect disk has another symmetry built into its geometry as well. If you rotate a perfect disk around its axis—like spinning a wheel—by any angle, it doesn't change.

It has the highest degree of rotational symmetry of any flat geometrical pattern. The total set of symmetries possessed by a perfect disk contains within it both these flipping and rotational symme-

---

[2] Once again, not including gravity.

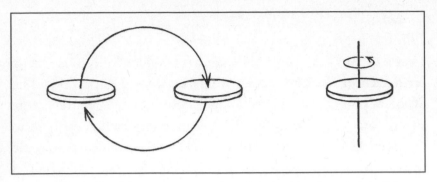

A perfectly flat disk possesses a number of symmetries. For example, it is the same regardless of which side is facing upward. Also, it does not change when rotated about its axis by any angle.

tries. In other words, the symmetry group that describes a disk contains within it multiple symmetries. More complex geometrical forms can, in many cases, have even more kinds of symmetry wrapped up inside of them.

Although the patterns behind the Standard Model—SU(3) × SU(2) × U(1)—are often talked about as if they were a single symmetry group, they are really more like three separate symmetry groups combined or stuck together. But just as the structure of a perfect disk possesses within it two different simple kinds of symmetry, the underlying unified theory behind the laws of our Universe may in fact be a single symmetry group that leads to the appearance of many subsymmetries—such as the three which make up the Standard Model. These symmetries are, however, quite intricate and complex. Any grand unified theory that contains all of the symmetries found in our world must be based upon a very large and complex symmetry group—something far richer and more multifaceted than a disk or any other simple geometric design. After all, such a theory must contain within its structure the very being of the strong, weak, and electromagnetic forces, as well as the patterns behind the quarks and leptons found in our world. Deciphering how all of these aspects of our world fit together into a single grand unified theory would appear to be a task of staggering difficulty. For a few years in the 1970s, however, the quest to unveil the grand unified theory behind our universe looked very close to being ful-filled.

THE PROCESS OF DISCOVERING how different aspects of nature are related through a unifying concept or principle is something that has repeated itself over and over again throughout the history of science. Indeed, it is at the very heart of nearly every great scientific accomplishment. Isaac Newton's greatest realization was that the force of gravity pulling objects downward toward Earth was the same aspect of nature that holds the planets in their orbits around the Sun. The great nineteenth-century physicist James Clerk Maxwell transformed our understanding of the Universe when he identified the phenomenon of magnetism to be nothing more than electricity in motion. Einstein's general theory of relativity explained why massive objects are both heavy and difficult to slow down. In other words, he unified the concepts of gravitational mass and inertial mass.

To those of us living in this modern scientific age, it is easy to take these remarkable insights into the workings of our world for granted. We are tempted to think, "Of course the force of gravity pulls us downward *and* holds the planets in their orbits," but this was very far from obvious in Newton's day. The fact that bolts of lightning had anything at all to do with why a compass points north was entirely unknown before the nineteenth century; in fact, such a connection may have seemed rather implausible. But a closer look at our universe can sometimes reveal a new set of patterns and a deeper understanding. If gravity pulls all masses toward one another, as it pulls objects toward Earth, then it must also pull planets toward the Sun. If an electric charge attracts and repels other charged objects, as we understand it to do, then we can also consider what effects the motion of an electric charge must bring about. Through this connection, magnetism is an inevitable consequence of electricity—just as planetary orbits follow necessarily from the laws of gravity. Seemingly separate and unrelated natural phenomena have been shown throughout history to be deeply and profoundly interconnected. Through the act of unification, multiple aspects of our world are brought together into a single conceptual framework. The search for unification is at the very heart of all science.

Grand unified theories follow in this same great intellectual tradition. They begin by positing that the various forms of matter— quarks and leptons—are fundamentally the same. Furthermore, they propose that the three forces of the Standard Model are not as independent as they appear, but instead are only seemingly different manifestations of the same grand unified force. If a grand unified theory is in fact behind the structure of our universe, then our world is much simpler than we realize. Although nature might look complex and diverse from our vantage point, all indications are that it was born of a simple pattern—and from a simple symmetry.

THROUGHOUT THE TWENTIETH CENTURY, many physicists tried and failed to build a theory that could encapsulate the combined laws of our universe. Most famously, Einstein spent much of his later career unsuccessfully searching for some kind of theory that would combine or unify the forces of electromagnetism and gravity. During his lifetime, however, the quest of building a unified theory was a wildly difficult one. There were simply too many pieces missing to make any progress on the puzzle of unification. When Einstein started his work on developing a unified theory, the only known particles were the proton, the electron, and the photon—not even the neutron, the positron, or the muon had been discovered yet. Electromagnetism and gravity were the only known forces. Very few pieces of the puzzle lay on the table before him. Many, many others were nowhere to be found.

As the decades passed, however, more and more of the missing puzzle pieces began to turn up—essential pieces, in particular the principle of gauge symmetry. With the discovery of the Standard Model, enough was finally understood about our world to place the goal of grand unification within reach. Most important, it had become clear that the three forces of the Standard Model each have a great deal in common—much more so than the forces of gravity and electromagnetism do. To begin with, they are each based upon the principle of gauge symmetry. In a sense, it was only natural to contemplate how these three gauge forces might be interconnected, or even unified into a single theoretical framework.

One of the most significant pioneers in the modern quest for a grand unified theory was a young physicist by the name of Howard Georgi. You might remember him from an appearance in the previous chapter, where I wrote about how he and Savas Dimopoulos proposed the first realistic supersymmetric model in 1981. Well, almost a decade before his foray into supersymmetry—indeed, well before he had even heard of supersymmetry—Georgi began to work on theories that could, from a single mathematical symmetry, give rise to the entire Standard Model. He had begun working toward building a grand unified theory.

After completing his doctorate at Yale in 1971, Georgi found himself in a research position at Harvard where he collaborated with Sheldon Glashow—one of the physicists behind the development and discovery of the Standard Model itself. It was a collaboration that paid off immensely. Over the next few years, they published a number of groundbreaking papers together. It was in one of these papers that the first modern grand unified theory was proposed.

But the work of Georgi and Glashow did not come out of thin

Howard Georgi and Sheldon Glashow (with bubble caption added by the physicist Alvaro DeRujula).
Photo provided by Howard Georgi.

air. A couple of years earlier, two physicists by the names of Jogesh Pati and Abdus Salam had written a pair of papers that made the seemingly outrageous suggestion that leptons and quarks might not be fundamentally different entities after all—a lepton might be just another kind or, more specifically, another color of quark. Georgi and Glashow began their quest for unification with this idea in mind, but took it much further. They imagined a theory in which the three forces of the Standard Model were not fundamentally different or separate from each other. In other words, their vision for our universe was that of a fully interconnected theory, in which all varieties of matter and force are manifestations of a singular form of force and matter—a fully unified picture of our world.

The crux of the Georgi-Glashow theory was that somehow the symmetries of the Standard Model are the products or components of some larger overarching symmetry. The symmetries of the Standard Model are, after all, not really one symmetry group as they are sometimes presented, but three. The SU(3), SU(2), and U(1) groups are essentially glued together to make up the Standard Model. Georgi and Glashow envisioned a single, but much larger, symmetry group that could include within it all of the symmetries of the Standard Model.

Einstein struggled for half of his lifetime to build a unified field theory of gravity and electromagnetism and still failed. When Georgi set out to build a grand unified theory, it took him all of one night. In the version of the story that I have been told, he sat down at his desk at home one evening in 1973. The first thing he tried was to build a theory using a symmetry group known as SO(10)—a much larger and more complex group than the SU(3), SU(2), or U(1) groups which make up the Standard Model. Amazingly, the entire structure of the Standard Model's symmetries is contained within the patterns of SO(10). After a glass of Scotch whiskey, Georgi realized that a smaller group inside of SO(10), called SU(5), also has all of the properties he was looking for. A few hours and Scotches later, he had formulated the first grand unified theory based on this symmetry.

As it turns out, there are only nine classes of symmetry groups that are large enough and complex enough to contain within them the Standard Model. Of these nine, Georgi and Glashow managed to find problems with eight of them; only those containing SU(5) survived. It was becoming increasingly clear that whatever symmetry the grand unified theory of our universe is based upon, it must contain the patterns of SU(5).

Recall that the group responsible for the strong force of the Standard Model—SU(3)—leads to the existence of eight force-carrying bosons, called gluons. Well, SU(5) works in the same way, but with a whopping twenty-four boson particles.[3] These twenty-four particles collectively make up all of the bosons of the Standard Model and

---

[3] As a side note, the number of bosons in a theory based on any SU(X) group is simply given by $X^2 - 1$.

then some. Eight of them are the strong force carrying gluons, while four others are the photon and the Z, W⁻, and W⁺ bosons. The other remaining twelve are colossally heavy particles known as X bosons—far too heavy to be observed in any plausible particle collider experiment. Together, these twenty-four particles constitute the singular force known as the grand unified force. The strong, weak, and electromagnetic forces are each parts of this unified force, much like magnetism and electricity are parts of the electromagnetic force. The three forces of the Standard Model are inseparable and interconnected aspects of the same underlying symmetry.

In some ways, Georgi and Glashow's grand unified theory might seem to make things more, rather than less, complicated. For example, twelve new bosons were introduced by the theory. But in other ways, the SU(5) GUT makes things considerably simpler. For one thing, the Standard Model has a number of parameters—nineteen to be specific—that are not determined by the theory itself. In other words, these quantities—sometimes known as fundamental constants, such as the electron mass, or the strengths of the various forces—simply have to be measured. The Standard Model does not predict their values. Ideally, we would like to have a more useful and powerful theory that could be used to predict the values of these quantities. Georgi and Glashow's grand unified theory, to some extent, does exactly that. For example, a quantity known as the Weinberg angle—which determines how much of the SU(2) and U(1) parts of the Standard Model contribute to each of the weak and electromagnetic forces—is predicted by the SU(5) GUT to be about 27.2 degrees. This is remarkably similar to the value that has been measured by experiments. The ultimate goal of grand unified theories—and science in general—is to predict as much about nature as possible. It is hard to overstate the significance of a theory that could predict the values of all—or at least most—of the fundamental constants found in nature.

BY THE END OF Howard Georgi's famous evening of Scotch drinking and GUT building, he realized something else about his SU(5) model—something that worried him a great deal. His theory implied that protons should decay. The atoms and molecules that make up so much of our world, it seems, should be unstable.

When quarks interact with gluons, the quarks' colors can be changed in the process. The W and Z bosons also can transform some of the quantum properties of the particles they come into contact with. The forces of the Standard Model only change the properties of particles in limited ways, however. Most important, the Standard Model bosons never transform a quark into a lepton, or a lepton into a quark. In the Standard Model, quarks and leptons are fundamentally distinct entities, with a line clearly separating them from one another. In grand unified theories, this is no longer the case. According to Georgi and Glashow's theory, the line between quarks and leptons is blurry.

Just as the exchange of a gluon can change a red quark into a blue quark, the exchange of an X boson in a grand unified theory can transform a quark into a lepton, or vice versa. What worried Georgi about this was that the quarks inside protons, through the exchange of an X boson, could be spontaneously transformed into electrons or positrons. As a result, his grand unified theory predicted that protons should disintegrate into positrons and pions. The chemistry of our world, however, appears to be very stable. If SU(5) is correct, Georgi wondered, "why isn't our world falling apart?"

At the time of Georgi and Glashow's work on grand unified theories, experiments sensitive to proton decay had been able to rule out the possibility that protons disintegrate more frequently than once every 10^{27} years or so, on average—a very long time, indeed. But was this long enough to be a problem for the Georgi-Glashow theory? When they calculated how long protons should survive before decaying, Georgi and Glashow predicted an average half-life of about 10^{30} to 10^{31} years—thousands of times longer than had ever been tested.

The reason that proton decay is so rare in grand unified theories is that the particles mediating this process are so amazingly heavy. Recall that the weak force is much weaker than the electromagnetic force because the W and Z bosons are very heavy—and thus spontaneously "pop out" from the vacuum of space very rarely. Well, the X bosons that mediate proton decay are *trillions* of times more massive than the W or the Z. As a result, the force they bring forth is incredibly feeble. So, although protons may decay through the GUT force, they almost never do. Georgi's initial concern that the proton's instability ruled out his theory turned out to be unfounded. In fact, the prediction of protons decaying provided an opportunity to test his theory. If more sensitive experiments could be built to search for proton decay, then perhaps his grand unified theory could be confirmed.

To test the stability of such long-lived protons, huge quantities of protons have to be observed at one time. After all, no experiment can watch an individual proton for 10^{30} years to see if it decays. One could devise a way, however, of observing 10^{30} or more protons at once for one year and see if any of them disintegrate. This is exactly what was needed to be done to test Georgi and Glashow's grand unified theory.

When physicists around the world heard about the predictions of the Georgi-Glashow model, they began to develop new experiments to search for evidence of proton decay. Huge underground experiments were built in mines in Minnesota, Italy, the Soviet Union, Utah, and elsewhere. Using thousands of tons of water, enormous numbers of protons were carefully observed for any sign of decay. Steadily, the minimum possible value of the proton's lifetime became longer and longer. Gradually, it moved from 10^{27} to 10^{28} years, and then on to 10^{30} and 10^{31} years. By the end of 1982, these experiments had ruled out the possibility that protons decay with a half-life shorter than about 10^{32} years—even longer than Georgi and Glashow's theory predicted it should be. Disappointingly, their grand unified theory did not hold the secrets to our universe after all.

ALTHOUGH THESE EXPERIMENTS REVEALED that the original model proposed by Georgi and Glashow is not manifest in nature, physicists' interest in grand unified theories only continued to grow. Despite the failures of their specific model, Georgi and Glashow had managed to show the world that plausible grand unified theories could be developed with some truly remarkable properties and features. To avoid the new and more stringent constraints from proton decay experiments, GUTs were developed using even bigger and more complex symmetries than SU(5). These symmetry groups each have built within them the SU(5) group, but also contain other patterns and structures beyond this symmetry. SU(5) alone was the simplest of all possible grand unified theories, but who said nature has to be so simple? The simplest of all possible physical laws does not need to be the one manifest in our world.

Over the decades since the first grand unified theory was proposed, physicists have learned a great deal about the nature of these theories. One of the most important of these insights has been the recognition of the role supersymmetry can play in grand unification. To understand this connection between supersymmetry and grand unified theories, one first has to understand how the three forces of the Standard Model—with three different strengths and characteristics—come forth as manifestations of a single grand unified force.

In grand unified theories, the grand unified force is brought into being when particles exchange any of the many bosons found in the theory—the twenty-four found in Georgi and Glashow's SU(5) model, for example. Together, these many bosons make up what we call the grand unified force. But why is this called one force? Isn't it really just several different forces that happen to emerge from the same symmetry? Well, yes and no. In the Standard Model, it takes eight gluons to bring into being the strong force. We call it a single force instead of eight forces because all of the gluons share certain properties. Among other things, they all interact with quarks and other gluons with the same strength.

But the different bosons of a grand unified theory don't seem to interact with the same strength. Gluons, for example, interact with quarks much more strongly than photons or W or Z bosons do—that is why, after all, we call it the strong force. So, if this is the case, how can we expect to combine these forces together into a single grand unified force? If gluons, photons, Ws, and Zs are all truly part of the same underlying force, then they must each interact with the same strength.

Physicists have come to understand, however, that the strengths that we observe forces to have are not necessarily the same strengths they have under all conditions. In particular, the three forces of the Standard Model interact with different strengths at different temperatures. At ultrahigh temperatures—or among particles with ultrahigh energies—the strong, electromagnetic, and weak forces of the Standard Model all behave very differently than they do in the environments where we study them. In particular, the strong force becomes gradually weaker at higher temperatures, while the electromagnetic and weak forces become stronger. If grand unification really is part of our world then, at some ultrahigh temperature, these three forces must all interact with the same strength. In essence, they—along with the effects of the X bosons—take on a single strength and act as a single force. In this way, the fact that the strong, weak, and electromagnetic forces appear to be so different to us is something of an illusion.

When grand unified theories were first being developed in the 1970s and 1980s, the way in which the three forces of the Standard Model evolved with temperature was only crudely known. This made it impossible to tell whether they really did come together with a single strength at a very high temperature. This all changed, however, with the running of the experiment known as LEP—the Large Electron-Positron Collider.

LEP was the premier particle accelerator and collider of its day. When it started its operation in 1989, it smashed electrons and positrons together with 45 GeV of energy each—enough to create copious numbers of Z bosons and other elusive particles. LEP studied the properties of these particles with dramatically greater precision than any experiment had done before or since. By studying these particles and their interactions, LEP was also able to determine how rapidly the strengths of the forces evolved with temperature. In hindsight, this is perhaps the single greatest legacy of the LEP experiment.

What LEP found was either disappointing or inspiring, depending on your point of view. The disappointing conclusion reached was that the three forces of the Standard Model do *not* appear to evolve to a single strength at a single temperature. By the temperature at which the strength of the strong force had evolved to a value similar to the electromagnetic and weak forces, those two forces had long since passed the point where their strengths matched each other. If the Standard Model is considered alone, its three forces do not appear to unify with one another.

But as I said, disappointment was not the only response to these measurements. LEP also revealed a new reason for excitement and inspiration. Although the observations of LEP might seem like a disaster for grand unified theories, they were far from it.

In the top frame, the strengths of the three forces of the Standard Model evolve to similar values at high temperatures, but never unify. In the bottom frame, the same thing is shown, but including the effect of the superpartner particles predicted by supersymmetry. In this case, the three forces take on a single strength at very high temperatures, making grand unification possible.

If any as-yet-unknown particles exist with masses between those of the Standard Model and the energy at which the three forces unify, they may affect how the three forces' strengths evolve. In particular, if our world is supersymmetric, then the superpartner particles will change how this occurs. When the particles predicted by supersymmetry are taken into account, the LEP data shows that the strengths of the three forces of the Standard Model ultimately evolve to almost exactly the same value—within about 1 percent of each other. LEP showed us that without supersymmetry, grand unification is likely to be impossible. Once supersymmetry is included, however, grand unification is nearly perfect.

In the years since this revelation, the measurements of the LEP experiment have been interpreted by many physicists as evidence for both grand unification and supersymmetry. Much as other aspects of our quantum world have revealed, supersymmetry holds an integral place in the quest for a grand unified theory. Remarkably, it appears that the unification of the matter and forces of the Standard Model requires the world to be supersymmetric. Everything, it seems, is starting to come together. More and more of the puzzle's pieces are sitting on the table right before us. Now all that remains is to finally discover supersymmetry once and for all. To do that, however, we are going to need one hell of a machine.

We are only going up to Sinai
to get the tablets. We have
no right to write the
commandments.
—*Carlo Rubbia*

|||||||||| **8** ||||||||||

The Hunt Begins

In the opening pages of this book, I tried to capture the amazing degree of excitement that the act of discovery can bring forth. To learn what has never before been known is an experience that sends shivers down the spines of the curious. Only slightly less intense than the sensation of discovery itself is the feeling of being on the path toward discovery. There is something profoundly human about this feeling. Our prehistoric ancestors must have been very excited in the act of killing their prey. But, well before the kill, they were deeply thrilled by the hunt.

The hunt for supersymmetry has been going on for decades and continues in full force today. Despite the many efforts—and the many scientists and engineers—behind these searches, the super-partner particles remain undiscovered. With the Large Hadron Collider—the LHC—beginning its operation, this is likely to change soon. But this chapter is not about what is likely to be the final hunt for supersymmetry; it is about the first.

The early 1980s were an exciting time for the physicists building and operating particle accelerators, and for the scientific community as a whole. In 1981, a new and more energetic particle accelerator—the Super Proton Synchrotron, or SPS—began its operation at the European CERN laboratory in Geneva, Switzerland. The SPS was an amazing machine. It accelerated protons and antiprotons around a giant ring, four miles in circumference, and smashed them together inside elaborate particle detectors. Each of the detectors was made up of more than two thousand tons of material and contained what was, at the time, perhaps the most sophisticated electronics in the world.

The two hallmark achievements of this experiment each came in 1983. In January of that year, the detectors of the SPS observed W bosons directly for the first time. In May, they detected Z bosons. With these two great accomplishments, any lingering doubts physicists might have had about the validity of the Standard Model were crushed. It was a great year for particle physics. It was a great year of discovery.

That being said, these achievements were not much of a surprise. Even before 1983, confidence in the Standard Model was very strong among particle physicists. It had, after all, been indirectly tested in a variety of ways, and was very successful. Although it was certainly important to confirm the Standard Model by observing the W and Z bosons directly, finding that these particles were present and that they had the properties they had been anticipated to have was not much of a shock to the scientific community.

Even the Nobel committee seemed unusually confident in the results coming from SPS. In 1984, it awarded two scientists from the SPS—Carlo Rubbia and Simon van der Meer—with the Nobel Prize in physics for the discovery of the W and the Z. This was a strikingly fast decision for the Nobel committee. By comparison, consider that it took sixteen years for the Noble committee to recognize Einstein for his prizewinning work. Dirac, Schrödinger, and Heisenberg each waited several years for their prizes—although not as long as

Einstein. In 1983, the year immediately preceding Rubia and van der Meer's own prize, the physics Nobel went to Subrahmanyan Chandrasekhar for work he did in the 1930s—nearly fifty years earlier. For the Nobel Prize in physics to be awarded for such a recent accomplishment was a virtually unprecedented event.

Following their discovery of the weak force bosons, the scientists running the SPS accelerator began to turn their attention to even more lofty goals. They had succeeded in their quest to validate the Standard Model. To accomplish this, they had built and operated the most energetic and sophisticated particle accelerator the world had ever seen. To follow such great successes, they would have to embark upon something truly spectacular—something that sought out entirely new kinds of physics. In 1984, they began a quest to do just that. They set out upon the first hunt for supersymmetry.

IT DIDN'T TAKE LONG for the physicists of the SPS to shift their attention from the W and the Z and begin searching for other new particles. Within a couple of weeks following the discovery of the Z, Carlo Rubbia—the head of the physicists operating one of the SPS detectors—began talking openly about other new kinds of physics they might soon find. Near the top of their list was the missing sixth quark of the Standard Model—the top quark. If it were light enough, it was thought that perhaps the top could show up in the future run of the SPS. Unfortunately, it never did. Today, we know that the top quark has a mass of about 171 GeV, making it by far the heaviest fundamental form of matter to be discovered so far. With such an enormous mass, the top quark was well beyond the reach of the SPS. (In fact, the top quark managed to evade detection for an entire decade. Eventually, it was discovered at the Fermi National Accelerator Laboratory outside Chicago in 1995.)

But the top quark was not the only kind of new particle that

Rubbia and his collaborators hoped to find with the SPS. Theoretical physicists had many ideas about what might show up in these experiments as they became more and more powerful. One of the most exciting of these many possibilities was that they might, for the first time, witness the particles associated with the theory of supersymmetry.

In the early 1980s, supersymmetry was for most particle physicists a new and very speculative idea. According to the records of the SPIRES archive at Stanford, only 107 scientific articles containing the words *supersymmetry* or *supersymmetric* in their title were published in 1980. Considering that there are many thousands of physicists around the world, this is quite a small number. But, although small, this number had been steadily increasing every year since supersymmetry had been introduced. As time went on, supersymmetry became more and more interesting to more and more physicists. In 1981, the 107 articles turned into 149, and then ballooned to 383 in 1982. By the middle of the decade, about 500 papers about supersymmetry were appearing in scientific journals each year.

Many factors contributed to supersymmetry's rise to prominence among particle physicists during this period. The introduction of the MSSM and the realization that supersymmetry provided an elegant solution to the hierarchy problem certainly played an important part. The ability of R-parity to stabilize the proton and to solve other problems was also becoming better appreciated. Furthermore, the lack of progress being made in other areas of particle physics—grand unified theories in particular—probably also contributed to supersymmetry's growing popularity. But in addition to these reasons, many physicists found themselves suddenly enthralled with the theory of supersymmetry because—for one fleeting moment—it looked like it may have finally been experimentally confirmed.

———

CRUDELY SPEAKING, ACCELERATOR EXPERIMENTS work by smashing together pairs of particles at incredible speeds and watching what comes out. The SPS was no exception. This experiment collided together huge numbers of protons and antiprotons, each with a few hundred GeV of energy. When a Z boson was created in one of these collisions, sometimes it would decay into an electron and a positron, or a muon and an antimuon, or a tau and an antitau, for example. Similarly, W bosons can decay into an electron, a muon, or a tau, along with a neutrino. The electrons, muons, and taus, along with their antimatter counterparts, can be detected as they travel away from the collision and out of the detector. Although neutrinos are essentially invisible, their presence can often be inferred by the energy they carry away—known as "missing energy." It was by looking for collisions with these kinds of features in the SPS data that the W and the Z were first discovered.

As physicists scoured through the data that was used to discover the W and the Z, they also began to find a handful of other strange and unexpected collisions. In addition to large amounts of missing energy, these events contained especially violent sprays of quarks and gluons, called "jets." At the time, it was thought that according to the Standard Model, these kinds of events shouldn't have appeared. For many of the physicists working on the experiment, it began to look like the SPS had once again discovered some kind of new phenomena. Maybe even a new type of particle.

The missing energy and jet events were especially exciting to the theoretical physicists advocating supersymmetry. If supersymmetry exists in nature, and the superpartners are fairly light, then collisions at the SPS should have been capable of occasionally producing pairs of supersymmetric particles—mostly squarks and gluinos. In a fraction of a blink of an eye, these particles would decay, often into a quark or a gluon accompanied by a neutralino. In the detectors of the SPS, the quarks and gluons would take the form of a jet. The neutralino would be invisible and be perceived as nothing but missing energy. According to the theorists, such occurrences should be rare. Squarks or gluinos should appear in about one collision out of a billion. But though rare, these events should occur.

Throughout the 1983 run of the SPS, a handful of jet events with missing energy were recorded, but the data was far from conclusive. Although the results were published, it was in a form that was unclear and difficult to interpret. Many physicists inside and outside the experiment were unconvinced by the findings. To resolve the questions these strange new events raised, more and better data would be needed. Fortunately, this is exactly what was about to be provided. For its 1984 run, the SPS experiment was upgraded considerably. In its new form, the accelerator would be capable of colliding particles together with more energy and at a faster rate. If the observed events were in fact the first evidence of supersymmetry, then the improved experiment would surely reveal this to the world.

WHILE THE SPS WAS shut down in preparation for its upgrade, the theorists working on supersymmetry were hard at work predicting what should be seen in the upcoming run if the jets were in fact the results of squarks or gluinos being produced. They concluded that in addition to more events with a jet and missing energy, another signal should begin to emerge once the energy of the accelerator was increased. These new events would be similar to those seen in the earlier run but with missing energy and two energetic jets—"dijets." To those who wondered whether the monojets, as the initial jet events came to be called, were indeed evidence of supersymmetry, the presence—or absence—of dijets would be the proof in the pudding.

As the upgraded run of the SPS began, the number of collisions that could be studied, and the energy of those collisions, was unlike anything that had ever been achieved before. Even more evidence for the W and the Z was accumulated, as well as more monojet events. The machine was operating fantastically.

One night in October of 1984, a handful of SPS scientists were monitoring the new events from the control room of one of the detectors. The room was some forty feet across, and contained more than sixty television monitors. Appearing on these various screens were numbers and images displaying the newly collected data. Some of the pictures resembled something out of the science fiction movie *Tron*. Billions and billions of collisions were taking place—far too many for the physicists to scrutinize individually. Instead, computers were programmed to identify the collisions that were particularly noteworthy. From time to time, an event would be flagged by the automated program and then looked at by hand. On that night in October, the computers identified one very interesting collision. The first dijet event had been spotted.

The first dijet—called event number 455—contained two very energetic jets, a less-energetic third jet, and a great deal of missing energy. It was exactly what they had been looking for, and exactly what the theorists studying supersymmetry had predicted. It was a beautiful event, very clean, and almost perfectly aligned to the detector.

Although it was not yet supposed to be made public, the news of the dijet event spread quickly. The physics rumor mill can at times be remarkably efficient, especially with news as exciting as this. But even as word of the event traveled, many scientists had their doubts. Perhaps the missing energy was not really missing but simply inadequately measured by the detector, thus faking the event. Or perhaps the ordinary, nonsupersymmetric processes weren't understood well enough—although dijet events were expected to be exceedingly rare in the context of the Standard Model, some wondered whether there might be ways to generate them without supersym-

metry that had not been fully taken into account. As Carl Sagan often said, "Extraordinary claims require extraordinary evidence." If
the SPS scientists wanted to claim that these events were in fact the
first signals of supersymmetry, they would need extraordinary evidence. They had clues. They needed a bulletproof case.

EVEN WITHOUT THE SQUARKS, gluinos, and other particles
predicted in the theory of supersymmetry, physicists knew that
events such as the observed monojets and dijets could occur, although only very rarely. For example, the missing energy in these
events might not be the result of a neutralino escaping the detector,
but could instead be an ordinary Standard Model neutrino. Similarly, until the physicists had crossed every "t" and dotted every "i"
in their analysis of the data, they could not be completely sure that
the observed jets weren't the product of ordinary quarks or gluons
produced by ordinary means. For example, instead of being generated by squarks or gluinos, there was the possibility that the monojets could be created through an ordinary process such as a Z boson
decaying to a neutrino and an antineutrino immediately after radiating a gluon. In some rare cases, the neutrino and antineutrino could
provide the missing energy of a monojet while the gluon could mimic
the energetic jet. Processes such as this could plausibly fake one of the
most sought after experimental signatures of supersymmetry. If the
monojets and dijets were to be thought of as conclusive evidence for
supersymmetry, great care would have to be taken to make sure they
weren't merely rare, but otherwise ordinary events.

Fortunately, there are ways to separate Standard Model processes
(like the one shown on page 150) from supersymmetric events. In
squark or gluino decays, for example, the jet that is produced is always
very energetic, whereas the gluon-jet in the Standard Model process
almost never is. Furthermore, quantum field theory can be used to

One of the ways a rare but otherwise ordinary Standard Model process could mimic a supersymmetric event. The process shown above could result in a monojet event, much like one expected from a squark or gluino decay into a quark or gluon (jet) and a neutralino (missing energy).

accurately predict just how often these Standard Model processes should occur, and just how often their jets will be energetic enough to mimic a supersymmetric event. When these calculations were first performed, they showed that there should have been less than one "fake event" seen from the class of processes shown in the figure. In comparison, six energetic monojets had been seen thus far.

So the scientists of the SPS could sleep well knowing that Z bosons decaying to neutrinos were not likely to be misleading them into thinking they had found the first evidence for supersymmetry. The problem was, however, that there were many other processes that needed to be considered as well. To make a claim of discovery, they would have to be sure that they had accounted for every one of the possible backgrounds, and every possible way that their signal could have been faked. Before they could claim to have discovered supersymmetry, or anything else, they had to be sure that they understood every detail of what went into their analysis.

In the first months of 1985, scientists from the SPS began giving talks at conferences describing the results of their new run. In these presentations, the world was officially informed that the monojets seen in the earlier run continued to appear and seemed to be authentic, and that dijets had also been found. All signs seemed to point to the existence of new physics—the most prominent candidate being supersymmetry. In the February 1985 issue of the journal *Nature,* an article appeared with the title "Hope Grows for Supersymmetry." That about summed up the sentiment of the time. To the world of particle physics, it seemed that supersymmetry was in the very act of being discovered. The only problem was that it wasn't.

IN THE FIRST MONTHS of 1985, excitement surrounding the search for supersymmetry at the SPS peaked. Rumors of powerful monojet and dijet events with lots of missing energy were rampant. There was hardly a particle physicist alive who was unaware of the exciting news. Science journalists began to write articles about the rumors, sometimes seeming to promote speculation to fact in the process. It was a time of great anticipation—and not nearly enough skepticism.

With hindsight, it is easy to say that some of the SPS physicists had been overeager to conclude that supersymmetry had been discovered. Compared to the experimental physicists running the SPS and its detectors, however, it was the theoretical physicists who were the most wildly enthusiastic. Several groups of particle theorists moved on from discussing whether the monojets and dijets could be generated by superpartners to working on detailed interpretations of the newly observed events in terms of supersymmetry. They worked out how heavy the squarks and gluinos needed to be in order to generate the SPS signals. Some argued that the squarks and gluinos each had to have masses in the ballpark of 40 GeV to account for the events. Others argued for much lighter gluinos and much heavier squarks. They debated and quarreled amongst themselves, and made predictions that could eventually be used to confirm or refute their various models of supersymmetry.

The feeling of impending discovery reached even greater intensity briefly that winter when yet another new signal appeared in the SPS data. This signal was seen in a number of the collisions that appeared to contain a decaying W boson. When these events were plotted on a graph, there was a clear and unexpected bumplike feature. Instead of the smooth distribution that was anticipated, this bump in the data prompted some to think that they were seeing evidence of yet another new kind of particle. In particular, such a bump could be the result of a supersymmetric particle such as a chargino decaying into an electron and a sneutrino, for example. Such a process would look a lot like a W boson decaying into an electron and a neutrino, but with a different mass and energy. In other words, it would look like a new feature in the distribution of W boson–like events.

Rubbia and the other physicists studying the bump didn't want the rest of the physics community to learn about this new signal until they had a chance to scrutinize it and understand it better themselves. They didn't want the wild rumors and speculation surrounding the monojet and dijet events to happen all over again. And they certainly didn't want the physicists working on the other SPS detector to learn about the signal and beat them to the discovery before they had a chance to go public themselves. Physics can be a very competitive business when the stakes are this high.

It is hard—some would say impossible—to contain such a rumor for long, but it was critical that they do so for as long as possible. Rubbia made his seriousness on this issue painfully clear to his collaborators: "I will deliberately try to follow any leak that comes back to me and go back to the origin: like the CIA or the KGB. All right?" Eventually, they would tell the world about the bump, but not until they were ready.

The secret, however, did not have to be kept under lock and key for very long. In order to study the bump feature more carefully, the SPS physicists designed and ran a computer program to calculate the distribution of events with an electron and missing energy that should come from ordinary Standard Model processes. The idea was to compare the results of the computer program to the data that they had collected. When the program was run, its output matched the SPS data almost perfectly—bump and all. The bump was not evidence of supersymmetry after all.

This all happened only a few days before a much anticipated particle physics conference in the tiny Italian village of Saint Vincent. In this hamlet, more than two hundred physicists gathered to hear about the new results from the SPS. What was the status of the monojets and dijets? What mass did the gluinos and squarks need to have to best fit the data? Were they finally ready to say for certain that supersymmetry had been discovered?

Rubbia did the best he could to speak at the conference without directly answering any of these questions. He listed the numbers of new monojets and dijets that had been recorded, but gave precious little information about the detailed properties of the events. He flashed pictures of some of the most energetic events but cautioned the audience that the results were preliminary and were not to be used by the theorists to discriminate between their various supersymmetry models.

As the meeting went on, it became increasingly clear that the claims of supersymmetry's discovery were built upon an unsteady foundation. Theorists specializing in quantum chromodynamics—the physics of quarks, gluons, and the strong force—presented the results of their detailed calculations. If you included accurately all of the Standard Model processes together, they argued, you could plausibly account for most, or even all, of the monojet events that had been observed at the SPS. Some of these events probably resulted from Z bosons decaying into neutrinos and a gluon. Others were likely collisions with a gluon and a W boson that decayed into a tau lepton and a neutrino. A handful of other events could have been the result of detector flaws. All in all, the monojets did not stand up well to the scrutiny they faced at the Saint Vincent conference.

And although the calculations regarding the dijets were even more difficult than for the monojets, these events, too, appeared to be plausibly explained in terms of ordinary Standard Model processes. Only a week before the conference, it had been thought that perhaps three separate signals of supersymmetry had been observed. But by the meeting's end, all confidence in this conclusion had evaporated. The rumors of supersymmetry's discovery were nothing more than a product of chance, circumstance, and a few overly eager scientists.

If it had been up to the physicists of the SPS to decide what form nature would take, then things would have surely turned out differently. But the laws of nature are for physicists to reveal, and not to select. As Carlo Rubbia once said, "We are only going up to Sinai to get the tablets. We have no right to write the commandments."

THERE WAS A GREAT deal of disappointment among physicists as what once had appeared to be evidence for supersymmetry dwindled away. But though it was not discovered by the SPS experiment, supersymmetry may still very well exist in nature. Mount Sinai is still being climbed. From today's vantage point, it's easy to see that the searches of the mid-1980s merely reached the first base camp. A great deal of progress has been made since, but evidence for supersymmetry remains elusive. The summit is, however, soon to be within reach. If supersymmetry indeed exists in nature, then the next particle accelerator—the Large Hadron Collider—is almost certain to discover it. Whatever the tablets have to say, this fantastic machine will reveal it to us.

> For every one billion particles of antimatter there
> were one billion and one particles of matter. And
> when the mutual annihilation was complete, one
> billionth remained—and that's our present universe.
>
> —*Albert Einstein*

> Where the telescope ends, the microscope begins.
> Which of the two has the grander view?
>
> —*Victor Hugo,* Les Misérables

|||||||||| **9** ||||||||||

Bang!

Fourteen billion years ago, our world began. From a singular point in space and time known as the Big Bang, the universe came forth and expanded to become the vast realm in which we exist. Perhaps no other kind of science has captured so many people's imagination and has inspired humanity's sense of wonderment as much as cosmology.

Many people experience a powerful sense of awe when they ponder the overwhelming immensity of our universe, its amazing origin and history. In my experience, however, I find that most people don't appreciate just how spectacular it truly is. When I give public lectures on cosmology, I am often asked questions such as "What caused the Big Bang to happen?" or "What happened just before the Big Bang?" These questions are very intriguing, but they don't have good answers. They don't have good answers because they are not valid questions—and, furthermore, they miss the point of the

Big Bang. To ask what happened before the Big Bang is like asking what is below the center of Earth or what is north of the North Pole. The Big Bang was not an event preceded by another. It was *the first event*. It was not the beginning of a period or of an era—it was the beginning of time and space themselves. Nothing happened before it because there was no before it—time before the Big Bang did not exist.

In many ways, the science of cosmology is a new frontier of thought which has only recently begun to be explored. Don't get me wrong: Modern cosmologists have learned a great deal about our universe's origin, history, and evolution. Only a century ago, however, there was no science of cosmology to speak of. Almost nothing was known about where our universe came from, or how it came to take the form we find it in. Questions such as these were limited to the speculations of philosophers and theologians. Science had very little to contribute.

The genius of Albert Einstein put a stark end to that state of affairs. With the introduction of his general theory of relativity in 1915, cosmology was suddenly thrust into the realm of science. In Einstein's theory, physicists found the tools they required to address a vast new range of questions. When our universe is contemplated through the lens of relativity, it becomes clear that space and time are not the simple and static backgrounds that they had long been thought of as. Contrary to what our intuition might lead us to believe, space and time can be stretched or warped and are naturally driven to change and evolve. According to Einstein's theory, space itself is expected to expand or contract with time.

In 1929, the American astronomer Edwin Hubble confirmed this aspect of relativity when he recognized that the most distant galaxies observed—millions of light-years away—are moving away from us. Furthermore, he noticed that the farther away from us a given galaxy is, the faster it appears to be receding from us. The distances between these objects and us are, in fact, growing because the space between them and us is growing. In other words, our universe is becoming larger; space is expanding.

Just as our universe will be larger in the future than it is today, it was also smaller in the past. Running the equations of Einstein's theory backward in time, we can trace our universe's history back to the Big Bang—about 14 billion years ago—and then watch it unfold through the formation of the first nuclei, atoms, stars, galaxies, and planets. This timeline of events has been confirmed by a broad array of detailed and precise cosmological observations and measurements. For example, according to the Big Bang theory, the nuclei of the lightest chemical elements—hydrogen, deuterium, helium, and lithium—were mostly forged in the incredible heat of the first few minutes of our universe's history. Remarkably, these chemicals are found to be present in our world in the abundances predicted by the theory. Furthermore, the Big Bang is predicted to have left behind a body of radiation that exists everywhere and fills all of space. This background radiation—called the cosmic microwave background—was detected for the first time in 1963, and has since been studied in incredible detail. Through these and other cosmological measurements, the Big Bang theory has been repeatedly confirmed and verified. These measurements have become so accurate in recent years that the past decade or so is often referred to as the "Age of precision cosmology." Through all of the scrutiny it has faced, the predictions of the Big Bang have been confirmed again and again. The successes of modern cosmology have been truly spectacular.

At first glance, it might seem that the subjects of this book—particle physics and supersymmetry—have little or nothing to do with the science of cosmology. Particle physicists study the behavior of matter and energy over the smallest scales—billionths of billionths of meters—whereas cosmologists study the universe as a whole. But despite this apparent distinction, these two fields of scientific inquiry are tightly interconnected and absolutely essential to each other. Shortly after the Big Bang, all of space was filled with particles that were very hot and possessed great quantities of energy. To understand how these particles behaved, cosmologists need to know what particle physicists have learned from accelerator experiments. In this way, particle accelerators can be used to study our universe's early history—the first fraction of a second to follow the Big Bang. Machines such as the Super Proton Synchrotron and the Large Hadron Collider not only reveal to us how matter and energy behave, but also help us to understand our universe's distant past. In this sense, particle accelerators are nothing less than time machines.

It also turns out that there may be an important role to be played by supersymmetry in the science of cosmology. Much as problems in the underlying structure of our world—the stability of the Higgs boson and the unification of the forces, for example— seem to point us toward the existence of supersymmetry, our current understanding of cosmology also has indications that supersymmetry may be called for. In particular, astronomers have discovered that most of the mass in our universe is made up of the as-of-yet unidentified, invisible substance known as "dark matter." In fact, only a small fraction of our universe's mass consists of the ordinary forms of matter with which we are familiar. In order to account for this missing majority of our world, elements beyond those found within the Standard Model are required. Many cosmologists think that the dark matter of our world might consist of the superpartner particles predicted by supersymmetry. If this is the case, then supersymmetric particles left over from the Big Bang fill the space throughout our universe and exist all around us.

THAT MOST OF THE matter in our universe is invisible—or dark—has been known by astronomers for decades. This invisible matter might have remained unnoticed if not for its rather conspicuous gravitational pull. Despite their appearances, galaxies and clusters of galaxies are not made up mostly of stars, but of dark matter. The dark matter's gravity causes these enormous structures to move and rotate differently than they would otherwise. By studying their motion, it is possible to determine just how massive galaxies and clusters of galaxies are—and to determine just how much of their mass is invisible or dark.

The forces that keep galaxies in motion are basically the same as those at play in our own solar system. When planets travel around the Sun, they are held in their orbits by the balance of the gravitational and centrifugal forces. The gravitational attraction of the Sun pulls planets toward it, while the centrifugal force pulls them away—like children are pulled away from a spinning merry-go-round. When these two forces are balanced, a stable orbit can persist.[1] The same thing holds true for stars orbiting around the center of a galaxy. Our Sun, for example, moves around the center of the Milky Way at a speed of about half a million miles per hour. This might seem fast until you consider that the center of our galaxy is about thirty thousand light-years away. This means that the time required for the Sun to make a single orbit around our galaxy—called a galactic year—is about 250 million years. The billions of stars that make up the Milky Way collectively pull our Sun toward our galaxy's

---

[1] Just as with planets and stars, the force of gravity pulls children and a merry-go-round together, although far too weakly to hold the child in a stable orbit. If gravity were a billion times or so stronger, however, it could then balance the centrifugal force and cause the child to spin around the merry-go-round without holding on—a stable orbit. Of course, if gravity were a billion times stronger, we would all be crushed by Earth's gravity (a billion g's of force!), but let's ignore that detail for the moment.

center, while its motion creates enough centrifugal force to resist this pull and keep it in a stable orbit. The balance between these two forces, however, depends on how fast the star is moving, and on how massive the galaxy is.

When the rotational speeds of stars around galaxies were measured, a surprising conclusion was reached. Many stars move far too quickly to be held in their stable orbits. Due to an imbalance of gravitational and centrifugal forces, they should fly away from their host galaxies and off into intergalactic space—but they don't. In order for the force of gravity to hold stars in their orbits, there must be much more mass in galaxies than is accounted for by stars and other visible materials. In other words, the gravitational pull of some invisible substance—dark matter—is holding our galaxy and other galaxies together. This dark matter makes up most of the matter in our universe.

The discovery of dark matter was not the first time that an otherwise hidden object was revealed by its gravitational pull. In the first decades of the nineteenth century, only seven of our solar system's eight planets had been discovered—Mercury, Venus, Earth, Mars, Jupiter, Saturn, and Uranus. The eighth and final planet,[2] Neptune, was discovered not by being seen directly with a telescope, but by its gravitational influence on the orbit of Uranus. Uranus's orbit around the Sun is slightly perturbed by the pull of the otherwise difficult to notice planet Neptune. It was by carefully watching Uranus that Neptune's existence was first deduced. Similarly, it was by carefully observing the orbits of stars around galaxies that dark matter was discovered. The history of science—as with other areas of history—often seems to repeat itself.

---

[2] I'd like to offer my sympathies to those mourning the loss of Pluto's status as a planet. On August 24, 2006, Pluto was demoted by the International Astronomical Union from full-fledged planet to dwarf planet (and it is not even the largest of those).

By the end of the 1970s, a consensus had been reached among astronomers that most of the mass in our universe is very dim or invisible—dark matter. Initially, most astronomers thought this missing mass was probably some collection of faint stars or planets—perhaps not completely invisible, but too dim to be easily detected with telescopes. Perhaps most of the dark matter, it was thought, could consist of large, Jupiter-like planets, or stars that had used up all of their nuclear fuel—called white dwarfs. The dark matter might also have been made up of more exotic objects that are produced when a star explodes as a supernova. Depending on the mass of an exploding star, it leaves behind either an ultradense object made up entirely of neutrons—a neutron star—or a black hole. Collectively, objects such as dwarf stars, planets, neutron stars, and black holes became known as massive compact halo objects—or MACHOs for short.

Motivated by the hypothesis that MACHOs make up the dark matter of our universe, methods to search for these objects were designed and carried out. Powerful telescopes scoured the sky for evidence of dark stars or planets. Although these searches succeeded in discovering some MACHOs, they found far fewer than they had expected—and far too few to constitute much of the dark matter. Whatever the dark matter is, these efforts revealed to us that it does not consist of faint stars, planets, or other dense and concentrated objects.

Today, astronomers think that our universe's dark matter almost certainly takes the form of some substance far more exotic than mere stars or planets. In particular, it seems very likely that it is made up of a gas of particles unlike anything we have ever witnessed here on Earth. These particles move slowly through our and other galaxies, almost never interacting with anything, earning them the title of weakly interacting massive particles—WIMPs. Galaxies themselves are imbedded in clouds of these WIMPs. The clouds hold themselves together by their mutual gravitational attraction, and hold stars in their orbits through their powerful gravitational pull. Stars may be the part of galaxies that we easily notice with our eyes, but they are in fact a very small piece of the total mass.

The precise nature of the particles that make up our universe's dark matter is still a mystery. It was once thought that ordinary neutrinos of the Standard Model could be the WIMPs that hold galaxies together, but computer simulations designed to study how galaxies form have since shown us that this cannot be the case. None of the particles contained in the Standard Model seem capable of acting as the dark matter. Whatever substance or substances make up the dark matter of our universe, it is a form of matter that has not yet been discovered.

ALTHOUGH THE NATURE OF dark matter remains stubbornly unknown, particle physicist have come up with a number of reasonably good guesses—or candidates, as they prefer to call them—for what this substance might consist of. Dozens upon dozens of hypothetical dark matter particles have been proposed throughout the pages of the scientific journals. For some of these candidates, there are quite compelling reasons to hypothesize their existence. Many others have far less convincing credentials. Among the most attractive are those candidates for dark matter that come from supersymmetry—most notably, the neutralino.[3]

In order for a type of particle to be a viable candidate for dark matter, it must have a few rather special characteristics. For one thing, it cannot interact in any way that would make its presence too

---

[3] You might recall that there are actually four different kinds of neutralino in the Minimal Supersymmetric Standard Model. When I talk about "the neutralino" in this chapter, I am referring to the lightest of these four particles.

obvious. Whatever makes up dark matter is elusive. It does not often disturb the atoms and molecules of our world. It does not radiate or reflect light. It is a sterile and quiet substance.

A second requirement is that dark matter particles must have been moving rather slowly—compared to the speed of light—during the period of time in which galaxies were forming. Astronomers call dark matter with this quality "cold." It is this requirement that rules out Standard Model neutrinos, for example, as candidates for dark matter. Neutrinos are simply too light and quick to have congealed together into the galaxies we see in our world. The main reason that neutrinos are so fast moving, or "hot," is that they are incredibly light—almost massless. Only if neutrinos were much heavier could they have formed galaxies in the way that dark matter has.

A third and final requirement is that dark matter particles must be stable. That dark matter still exists today tells us that whatever it consists of does not significantly decay over the age of the universe. Given that our universe has been around for a very, very long time, this is a very stringent requirement.

Remarkably, supersymmetry readily provides us with a particle that satisfies all three of these requirements. Take a look again at the table of superpartner particles on page 113. Most of these species of particles can immediately be ruled out by the first of the three requirements I described above. In fact, any particle with either electric charge or color will interact far too much with ordinary matter to be the dark matter of our universe. This excludes selectrons, smuons, staus, down squarks, up squarks, strange squarks, charm squarks, sbottoms, stops, gluinos, and charginos as dark matter candidates. Only the neutralinos and sneutrinos remain as possibilities. It even turns out that sneutrinos, despite their lack of electric charge and color, would interact a bit too much to have gone undetected if they were in fact the dark matter. Of all of the new particles introduced by supersymmetry, the neutralinos stand out as our best candidates for dark matter.

But do neutralinos satisfy the other two requirements required of a viable dark matter candidate? Although we don't know exactly how heavy neutralinos are, they are likely to be quite heavy—assuredly much, much heavier than neutrinos. If these particles exist, they are almost certainly sufficiently heavy and slow to have formed the galaxies that are observed in our universe, thus easily satisfying the second requirement.

Constructing a theory with a particle that is both heavy and stable is not generally an easy thing to do. After all, most known varieties of heavy particles are far from stable. W and Z bosons exist for less than a billionth of a billionth of a billionth of a second, on average, before decaying. Top quarks live even shorter lives. The heaviest known particle that does not decay—at least not often enough to be detected—is the proton. Similarly, most of the heavy particles hypothesized by theoretical physicists are also very unstable. Stability is, perhaps, the most difficult requirement of the three to satisfy. If a heavy particle is to be a viable candidate for dark matter, it must somehow be prevented from decaying over cosmological lengths of time. Some new physical law or principle must forbid its decay. Without such a law, dark matter would have dissolved long ago.

Once again, supersymmetry provides a natural solution to this challenge. Recall from chapter 6 that in some early supersymmetric theories, protons were predicted to decay at an unacceptably fast rate—taking between a fraction of a second and a year before disintegrating. In order to prevent this disastrous process from occurring—or at least to make it extremely rare—the additional symmetry of R-parity conservation had to be introduced. This symmetry requires that superpartner particles are only created and destroyed in even numbers, such as pairs. An isolated superpartner can therefore never be produced in a collider experiment, for example, because one is an odd number. Similarly, R-parity conservation ensures that an isolated superpartner particle will never decay into ordinary Standard Model particles. Unless a superpartner particle encounters another superpartner, possibly leading to their mutual destruction, it will survive forever.

In the years that have passed since the realization that neutralinos could be sufficiently elusive, cold, and stable to make up our universe's dark matter, many experiments have been conducted to search for the presence of such particles. So far, no such detection has been made, but even this lack of evidence has taught us a great deal. For example, it was these kinds of experiment that taught us that sneutrinos do not constitute the dark matter of our universe. If they did, these experiments would have detected them long ago. Although many other dark matter candidates have been proposed, the lightest neutralino has been studied in greater detail and by more physicists than any other and remains our single best guess as to this elusive substance's identity. Although we cannot yet be sure, many physicists consider it likely that neutralinos constitute the bulk of the dark matter present in our world.

TO UNDERSTAND HOW OUR world came to contain the varieties of matter found within it today, one must look back to the first moments of our universe's history. In these instants can be found the origin of the protons, neutrons, and electrons that make up your body, Earth, the air you breathe, and all of the objects around you. Nearly all of the matter in our world was created in these very first fractions of a second following the Big Bang. The dark matter of our universe is no exception.

In the first instants after the Big Bang, our universe was unimaginably dense and unimaginably hot. So much energy was present that all forms of matter and antimatter were easily produced and

existed in vast quantities throughout all of space. It didn't matter
how heavy the various types of particles are, they could all be created effortlessly—there was plenty of energy to go around. The very
heavy top quarks and anti-top quarks were as common and plentiful
as the much lighter electrons and positrons. Higgs bosons were not
much rarer than up or down quarks. It was a democratic and egali-
tarian age in which all species of particles existed as equals. It was
a very different time and, in many ways, a very different world.

In this ultrahot state, the entire universe was filled with a dense
soup of particles, constantly colliding and interacting. These collisions frequently produced pairs of particles—electron-positron or
quark-antiquark pairs, for example. But at the same time, just as
these particle pairs were being created, collisions of other matter
and antimatter particles were destroying each other through the act
of particle-antiparticle annihilation. For each type of matter and
antimatter, there was a natural equilibrium between the ever-present
processes of particle creation and annihilation. It was like a perpet-
ual war, with the same number of casualties every day as new babies
born. It was a violent and turbulent time.

The all-particle soup that existed in the first moments after the
Big Bang did not persist for long, however. As the universe expanded
and cooled, the natural equilibrium between particle-antiparticle
creation and annihilation was broken. In a cooler universe, the
heavier species of particles could no longer be freely created; there
was simply no longer enough energy present to do so. Particle annihilations, in contrast, continued unimpeded. As the equilibrium between creation and annihilation came to an end, heavy particles such
as top quarks and W bosons became more and more rare until they
became practically nonexistent. By one trillionth of a second after the
Big Bang, the vast majority of top quarks, W and Z bosons, bottom
quarks, taus, and charm quarks, along with their antimatter counterparts, had disappeared.

But even as our universe cooled, bringing its fledgling democracy to a crashing halt, not every form of matter was wiped out. We see protons, neutrons, and electrons all around us. Somehow, these particles managed to survive the process of rampant particle annihilation that occurred in the early stages of our universe's history. These particles were spared. Their survival is due to the slightly broken symmetry between matter and antimatter in our world.

As the universe cooled, matter and antimatter particles annihilated each other in enormous quantities. At some point, however, this process came to an end. If it had not, there would be no matter left in our world today. Like an overaggressive and unregulated logging company that cuts down every last tree, the process of matter-antimatter annihilation came to an end because the antimatter present in the early universe simply ran out.

Immediately after the Big Bang, there were almost equal amounts of matter and antimatter in the universe—almost, but not exactly. For every ten billion particles of antimatter present, there were about ten billion and one particles of matter. After annihilations had proceeded for a brief moment, ten billion turned into a million and then a thousand and so on. In the end, the antimatter particles had all been destroyed. But although there was no antimatter left, a tiny residual of matter particles remained.

For there to be the quantity of protons, neutrons, and electrons left in our universe that we observe today, there must have been some tiny difference between the number of matter and antimatter particles that were present in the first moment of the Big Bang. For some reason, ten billion and one matter particles were created in the early universe for every ten billion particles of antimatter. The reason for this tiny—but crucial—asymmetry between matter and antimatter in our universe is unknown. It probably has something to do with slight differences in how particles of matter and antimatter interact. Particle physicists have constructed many theories of how such interactions could produce the matter-antimatter asymmetry, but which theory is correct is still a very open and very intriguing question. Regardless of what exactly caused this matter-antimatter asymmetry to exist, the atoms and molecules that we see in our world today are here because our Universe's laws are only almost perfectly symmetric. A more perfect symmetry between matter and antimatter would have rendered our Universe uninhabitable.

AS THE FIRST INSTANTS of our Universe's history passed, matter began to take increasingly familiar forms. A few minutes after the Big Bang, protons and neutrons began to bind together for the first time to form the nuclei of the lightest types of atoms—hydrogen, helium, and lithium. Although it would be hundreds of thousands of years before nuclei would combine with electrons to form the first complete atoms, it was in those first minutes that most of the light chemical nuclei in our world today were forged. But the ultrahot particle soup of the Big Bang did not give birth to all of the periodic table as we know it—only the lightest few elements came into being in this way. For the origin of the many heavier species of atomic nuclei, we have to turn to the engines of nuclear fusion—the cores of stars.

A few hundred million years after the Big Bang, clouds of hydrogen and helium gas began to collapse under the force of their own gravity to form the first stars. Under the incredible pressure of their own weight, the cores of these giant spheres of gas became very, very hot. So hot, in fact, that light atomic nuclei began to bind together through the process of nuclear fusion to form heavier nuclei. This process takes place in all stars, and is the source of their energy—and what causes them to shine. Depending on how massive a star is, the route through which it undergoes this process can vary. For smaller stars, hydrogen nuclei combine to form those of deuterium and helium through a mechanism called the proton-proton chain. In heavier stars, another process called the CNO cycle works to build up heavier elements. If a star's core is sufficiently hot, it will fuse helium nuclei together to form carbon, which in turn combines through a process called carbon burning to form magnesium, oxygen, and neon. The nuclei in the cores of heavy stars continue to be built up into larger and larger elements until they are mostly iron. Although our Sun will never reach the temperatures needed to fuse nuclei together into iron, by the time its life is over, it will have transformed most of the hydrogen and helium found in it today into carbon and oxygen.

The products of a star's nuclear burning have their limits, however. Although chemical elements as heavy as iron can be formed through this process, the cores of stars are not hot enough to manufacture many of the heaviest members of the periodic table. And although far less common than the lighter species, we do see these heavier elements in our world—lead, uranium, and gold for example. These substances do not owe their existence to the life of a star—but instead to a star's death.

When a star runs out of nuclei that it can fuse together into heavier chemicals—when it runs out of nuclear fuel—it becomes unable to produce the pressure needed to sustain itself, leading it to suddenly collapse under the force of its own gravity. In about 5 billion years, this will happen to our Sun, causing it to shrink and cool into a white dwarf star. But not all stars quietly retire as a dwarf. Others experience a far more dramatic death. In particular, when heavier stars collapse, their entire mass can be compressed into an absolutely tiny volume that then rebounds in a colossal explosion—a supernova.

In the fleeting moments of stellar collapse, the star's core can be so hot that all of the elements of the periodic table can be produced—uranium, radon, gold, and so on. Most of the heavy

TYPE OF MATTER	MECHANISM/TIME
Lightest chemical nuclei (deuterium, helium, lithium)	Protons and neutrons combine throughout space, approximately three minutes after the Big Bang
The first atoms	Protons and electrons combine throughout space, approximately four hundred thousand years after the Big Bang
Intermediate chemical nuclei (up to iron)	Nuclear fusion in the cores of stars
Heavy chemical nuclei (heavier than iron)	Supernovae explosions

Table 7. How and when the various nuclei and atoms in our world came to be formed.

chemicals in our universe owe their existence to these spectacular events. The lead in an old pipe, and the gold or silver in your ring, was born in one of these exploding stars.

Unlike atomic nuclei, dark matter was most likely not forged in the cores of stars or in supernovae explosions. It was not even formed along with the lightest nuclei a few minutes after the Big Bang. Instead, it was likely created even earlier, in the first fraction of a second of our universe's history. In this first flash of time, dark matter particles were plentiful and ever present. As the universe expanded and cooled, the other heavy forms of matter were destroyed, but dark matter survived. Just as dark matter's feeble interactions make it hard for us to see or experience it, they also make it difficult to destroy. Because of this property, dark matter is especially well suited to survive the ravages of the Big Bang.

Although most of the dark matter that existed in the early universe was destroyed along with the top quarks, W bosons, and other heavy particles, a small fraction was not. Those dark matter particles that managed to escape the fate of the other heavy particles in our universe's first instants occupy all of space and are all around us today. There is far more mass in our universe in the form of dark matter than in protons, atoms, neutrinos, and all other kinds of matter combined. If the dark matter particles are indeed neutralinos, then supersymmetry is not only imprinted upon the blueprint or structure of our universe itself, but also physically surrounds us on Earth and throughout all of space. It also provides us with another way of studying supersymmetry. Neutralino dark matter particles may be all around us, just crying out to be detected, observed, and studied. Although very elusive, such particles are not entirely imperceptible. To catch a glimpse of such a subtle substance, however, will require us to look with very new kinds of eyes.

IN THE SPRING OF 2007, I was part of a group of physicists who organized a conference at the Fermi National Accelerator Laboratory focusing on efforts to learn about dark matter. Scientists came to the meeting from all around the world to give and attend presentations about a wide range of experiments being developed to search for dark matter particles.

Many scientists at the meeting spoke about particle colliders, such as the upcoming Large Hadron Collider, which may actually be capable of manufacturing particles of dark matter. If dark matter in fact consists of neutralinos, then it is very likely that this new machine will tell us a great deal about this substance. But other, very different efforts to discover dark matter's identity—whether neutralinos or as some other form—were also discussed. In underground mines, miles beneath Earth's surface, experiments are being carried out in an effort to detect directly dark matter particles from space. Dark matter particles interact with ordinary matter so feebly that they pass through Earth without even noticing it is there. These particles, passing through our galactic neighborhood, can on occasion pass through Earth and into one of these mines, where there is a very small chance that they will collide and rebound with the detectors of one of these experiments. By operating deep underground, the detectors are protected from many of the ordinary kinds of particles that could be potentially confused with a signal of dark matter.

Furthermore, remarkable new kinds of telescopes have been built to search for the presence of dark matter in our galaxy. When two particles of dark matter collide, they can annihilate each another, becoming transformed into very energetic forms of matter—electron-positrons, protons-antiprotons, neutrinos and photons. Astronomers are currently searching for each of these forms of dark matter by-products.

The telescopes that are used in these searches are not like the telescopes you might imagine. To search for neutrinos from dark matter annihilations, for example, electronic detectors have been buried throughout a full cubic kilometer of ice at the South Pole. This giant cube of ice—called "IceCube"—is the world's largest neutrino telescope. To search for energetic photons and particles of antimatter, sophisticated detectors have been deployed on satellites orbiting Earth, and on high-altitude balloons circling the Antarctic continent.

There was an overwhelming sense of optimism that seemed to pervade the scientists attending the dark matter conference at Fermilab. This impression has really stayed with me ever since. When I started studying dark matter about eight years ago, the prospects of these experiments seemed remote. Today, the experimental technologies are developing to a point at which a discovery is becoming realistically attainable.

An unofficial survey taken at the conference asked whether those in attendance thought that dark matter would be detected—either at a particle collider, in an underground laboratory, or with the new generation of telescopes—within the next five years. When this question was raised, a staggering fraction of the hands went up—more than half, including my own. If these scientists are correct, then dark matter's identity will not be hidden from us for much longer. The question then becomes, which of these many experiments will reveal it first? The race is under way, and the stakes are high. As I've said many times before, it is a very exciting time to be a scientist.

> We are going to make that particle, or we
> are going to show that it doesn't exist.
>
> —*Jos Engelen*

> Our judge is not God or governments, but
> nature. If we make a mistake,
> nature will not hesitate to punish us.
>
> —*Jim Virdee*

||||||||||| **10** |||||||||||

The Machine

I f you ever get a chance to visit the Fermi National Accelerator Laboratory in Illinois, take a few moments to drive around the main parking lot. Interspersed throughout it, you will find cars with a wide range of physics-themed license plates—NUTRINO 3, TEV, QUARK 1, KAON, CQUARK 1, TOP QURK, HIGGS, and so on. Many of these plates are testaments to their owners' careers—physicists who have spent much of their lives hunting for or studying the properties of particles such as neutrinos, quarks, kaons, or the Higgs boson. These days, Fermilab's scientists hope that by smashing together pieces of matter with tera-electron-volts—TeV—of energy they will be able to create new forms of matter, such as supersymmetric particles.[1] In the prairie of Chicago's western

---

[1] In case you are keeping track of the metric prefixes, a TeV (tera-electron-volt) of energy is equal to 1 trillion electron volts, or one thousand GeV (giga-electron-volts).

suburbs, these scientists have built an experiment to explore the quantum nature of our world at these vast energies. This experiment is Fermilab's massive particle accelerator, called the Tevatron.

For the last two decades, the Tevatron has been the world's largest and most energetic particle accelerator. Using ultrapowerful superconducting magnets, this machine propels two beams of particles—one of protons, the other of antiprotons—around an underground ring four miles in circumference. As these particles are flung around this colossal circle, they are accelerated up to incredible speeds and energies. By the time that the two particle beams are collided into one another, they are moving at a speed greater than 99.9999 percent of the speed of light. When the two beams collide, so much energy is present in one place at one time that new kinds of matter can be formed and studied. So far, the Tevatron's single greatest achievement was its discovery of the top quark in 1995. Before the Tevatron, no experiment was able to collide particles together with enough energy to create top quarks. No one else even came close.

Colliding particles together may seem like a very peculiar method for studying the types of matter that exist in our world. It's hard to imagine an automotive engineer driving a car into a wall—or indeed into another car—in order to better understand how its carburetor or driveshaft works. But many phenomena in our universe only take place when a great deal of energy is present. Similarly, many forms of matter are only manifest when energy is plentiful and concentrated. We know of no better way of getting more energy into one point in space than to accelerate particles and smash them together as fast as possible.

The reason that energy is so closely tied to an experiment's ability to study new forms of matter comes down to Einstein's equation, $E=mc^2$. As I have said before, mass is just a form of energy. In fact, mass is something like a storage battery of energy. Among other things, this means that energy can sometimes be converted into—or stored as—mass. Mass can also sometimes be converted into energy. Nuclear power and nuclear weapons, for example, work by converting small amounts of mass into energy—lots of energy. In these processes, neutrons are transformed into a proton, an electron, and an antineutrino. All together, these end products are about 0.08 percent lighter than their parent neutron. The tiny bit of mass that is lost in this process becomes energy. In a nuclear chain reaction, this process is repeated, converting an enormous number of neutrons into protons, electrons, and neutrinos. With an enormous number of such conversions, an enormous amount of energy is released.

Particle colliders also make use of Einstein's relationship between mass and energy, but in the opposite direction. If you try to create a neutron from a stationary proton and electron, you will fail. Because neutrons are slightly heavier than the combined mass of these other particles, a little extra energy is needed if this conversion is to take place. In other words, the proton or the electron have to be moving in order to create a neutron. For an electron to strike a stationary proton with enough energy to potentially create a neutron, it would have to be traveling at just over 90 percent of the speed of light. This might sound like a lot, but it corresponds to about a thousandth of a GeV of extra energy—a million times less than the level of energy reached by the Tevatron.

The forms of matter that modern particle collider experiments most aggressively seek to discover—the Higgs boson, superpartner particles, and others—are, however, much, much heavier than neutrons. To create such very heavy particles, very large amounts of energy are needed. The simplest way to get more energy into a collision of particles is to accelerate the particle beams to the highest possible speeds—more than 99.9999 percent of the speed of light in the case of the Tevatron. Accelerating the beams to greater speeds means that more energy will be present when they are collided together. More energy means that heavier forms of matter can potentially be created and discovered. In this sense, an accelerator physicist is like an automotive engineer who finds that driving together two compact cars with their pedals to the metal can—on occasion—cause two Abrams tanks to fly out of the collision. This may sound wildly impossible but, to the particle physicist, this is simply how strange the structure of our quantum world is.

Although it has had an incredible run, the Tevatron's reign as the world's most energetic particle accelerator is coming to an end. The Large Hadron Collider—the LHC—is about to take possession of this crown. In fact, by the time this book goes to press, the first collisions at the LHC will likely have taken place. This spectacular new machine—superior in nearly every respect to its predecessors—is the experiment an entire generation of particle physicists has been waiting eagerly for. For thousands of particle physicists around the world, the day the first LHC results are announced will feel like Christmas to a ten-year-old. We have been waiting a long time for this—and we are expecting to find a very nicely stuffed stocking.

The potential for the discovery of new particles and new laws of physics at the LHC is overwhelming. If the Higgs boson exists, as the vast majority of particle physicists expect it does, this experiment will find it. If new gauge bosons, or new quarks, or other new forms of matter lie hidden, no other machine has ever had a better chance of revealing them. If supersymmetry is imprinted on the blueprint of our universe, it will almost certainly be unable to hide from the Large Hadron Collider.

COLLIDING PARTICLES TOGETHER IN order to better understand them may seem like a strange approach, but it is not by any means a new one. Over the past century, variations of this technique have been used again and again by those exploring the quantum nature of our world. Although modern colossal and sophisticated particle colliders such as the Tevatron and the LHC would have been hard to imagine a century ago, the method of propelling particles into one another to shed light on their nature had already been conceived.

One hundred years ago, scientists were just beginning to learn about atoms. Although the notion that matter might consist of tiny, indivisible pieces goes back to ancient times, it wasn't until the twentieth century that the atomic nature of matter was empirically tested and confirmed for the first time. In 1905, Albert Einstein's work studying the motion of microscopic grains—known as Brownian motion—provided the final proof that atoms and molecules exist. With that accomplishment—one of Einstein's several revolutionary and Earth-shattering achievements in that year—the last nail was driven into the coffin of the atom skeptics. Atoms do indeed exist.

But despite the confirmation of their existence, very little else was known about the atoms that make up so much of the matter in our world. What they consist of, how they are structured, and what laws govern their behavior were still unknown. The best guess of scientists studying atoms at the time was that they were probably made up of positively and negatively charged parts—what we now know to be protons and electrons. If these internal parts of atoms—such as the electrons—were moving too quickly, however, it was thought that they would radiate energy, causing the atoms to collapse. To keep the electrons moving slowly, the physicist J. J. Thomson hypothesized that they were embedded in a positively charged atomic medium—something like a thick pudding of positive electric charge. For this reason, Thomson's proposal was called the "plum pudding" model of the atom—electron plums spread out through a pudding of positive charge.

In his laboratory in Cambridge, England, the physicist Ernest Rutherford decided to put Thomson's plum pudding model to the test. Using an electric field, Rutherford and his colleagues Hans

Geiger and Ernest Marsden were able to accelerate nuclei of helium atoms—known then as alpha particles—toward a very thin foil of gold. The alpha particles were, for the most part, unscathed by the wimpy foil—being less than a thousandth of a millimeter thick, it was hardly noticeable to the particles passing through it. The alpha particles were like bullets passing through a thin layer of sand—or through a plum pudding.

If that were all that Rutherford's gold foil experiment had detected, it would be little more than a footnote in the history of science. But Rutherford's findings are no mere footnote—you can look them up in almost any high school chemistry textbook. The essential feature that Rutherford and his collaborators discovered was not how the majority of the alpha particles behaved, but how the very rare and exceptional ones did. Out of every twenty thousand or so alpha particles that passed through the gold foil, one bounced back, having been deflected by more than ninety degrees. Rutherford was shocked, saying, "It was quite the most incredible event that has ever happened in my life. It was almost as incredible as if you fired a 15-inch shell at a piece of tissue paper and it came back and hit you!"

No plum-pudding-like atom could possibly lead to this kind of behavior. The deflection of the alpha particles proved that there were compact parts inside atoms—concentrated cores of electric charge. Although most of the alpha particles traveled right through the gold atoms that made up the foil, a few smacked right into an atom's nucleus and bounced back. In essence, by observing these deflections, Rutherford and his colleagues had discovered atomic nuclei. Atoms are nothing like the blobs of pudding Thomson and others had imagined. Instead, they are tiny concentrations of mass and positive charge—the nucleus—surrounded by a vast cloud of orbiting electrons. Atomic nuclei are absolutely minuscule—ten thousand to one hundred thousand times smaller than the atoms

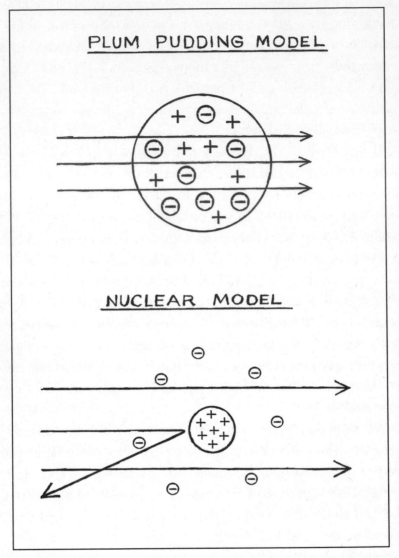

According to the plum pudding model of the atom, alpha particles shot through the gold foil should pass with only a tiny degree of deflection. Rutherford's experiment, however, found that about one in twenty thousand of the alpha particles were deflected backward. This was the first evidence that atoms contain a very concentrated nucleus.

they reside in. Rutherford's early collider experiment showed that atoms were very different than they had been thought to be. The strangeness of the quantum world had begun to rear its perplexing head.

DANIEL H. BURNHAM, THE great Chicago architect and master-mind behind the World's Fair of 1893, once said, "Make no little plans; they have no magic to stir men's blood." It would appear that the designers of the Large Hadron Collider have taken these words to heart. In my opinion, no more incredible machine has ever been created by humankind.

A few hundred feet beneath the city of Geneva, Switzerland, and the neighboring region of France is the circular tunnel—seventeen miles in circumference—around which the LHC accelerates protons. These protons are propelled by some of the world's most powerful magnets. When I say magnets, don't think of ordinary bar magnets, or something you might stick on your refrigerator. These magnets are huge—the biggest are about forty feet long and weigh 20 tons. In total, almost 10,000 magnets are required to keep the protons moving along their course around the track of the LHC, speeding them up as they go. These magnets are superconducting, so need to be kept at ultracold temperatures. This is done by using 128 tons of liquid helium, which has a temperature of less than 2 degrees above absolute zero—or 456.25 degrees below zero Fahrenheit.

When the LHC turns on, beams of protons will travel in both clockwise and counterclockwise directions around the underground ring at a speed of about 99.999999 percent of the speed of light. Once at this speed, a proton will make more than eleven thousand trips around the seventeen-mile ring every second. The beams can be kept moving around the LHC ring for hours. Over that length of time, and at nearly the speed of light, a proton can travel billions of miles—more than the distance from Earth to Pluto.

The protons moving around the LHC ring are grouped together in bunches, with about 100 billion protons in each bunch. Every second, about 30 million proton bunches are put on a collision course with a proton bunch from the other beam. Each time two bunches are fired into each other, about twenty pairs of protons are expected to collide with one another—leading to a whopping 600 million collisions per second to be observed, recorded, and studied. When I think about the staggering engineering challenges that go into a project like this, it makes the Moon landing seem like mere child's play.

The beams are crossed and collisions take place at four locations around the LHC ring. Around each of these four locations are the detectors of the LHC. The largest of the four detectors is somewhat ironically named the Compact Muon Solenoid, or CMS. As far as I am concerned, there is nothing compact about this colossal device. Consisting of over twelve thousand tons of electronics and metal, the cylinder-shaped CMS extends for about seventy feet and is more than fifty feet wide. As protons are smashed into each other in the heart of the CMS detector, they are ripped apart into a violent spray of particles, spewing out in all directions. The various kinds of particles produced travel into and through the various components and layers of CMS. From the imprint of these particles, the detector attempts to reconstruct what kind of interaction took place, and what kind of matter was present in the fleeting fraction of a second following each collision. Given that hundreds of millions of such collisions will take place in each of the detectors every second, simply recording all of the necessary data can be an overwhelming challenge. Each year it is running, the LHC is expected to record enough data to fill 3 million DVDs.

One of the Large Hadron Collider's four detectors,
the Compact Muon Solenoid.
Photo credit Peter Reid and SCI-FUN.

BUILDING THE LHC IS certainly one of the greatest and most challenging engineering endeavors to have ever been carried out. Not including the time for planning and design, it took a decade and a half—and about $8 billion—to construct. The collaboration of scientists and engineers who run the CMS detector consist of more than 2,500 scientists from thirty-seven countries. Another 3,500 people run the other three detectors at the LHC.

But as staggering as the technology of the LHC is, to a particle physicist, the machine itself is not nearly as interesting as the science it is designed to uncover. After all, the LHC wasn't built to merely be an incredible technological marvel—it was built to be a discovery machine.

To many particle physicists, the single most important task of the LHC is to finally discover the last missing piece of the Standard Model—the Higgs boson. It is through interactions with this very special particle—the "God Particle," as it is sometimes called—that the other fundamental particles are thought to attain their masses, transforming our world into a very different place than it would have been otherwise. The Higgs boson is absolutely essential to our understanding of our Universe.

Fortunately, the LHC is a machine extremely well suited for Higgs hunting. That is not to say that Higgs hunting is easy—it is certainly not. The LHC, however, has been designed to be particularly good at this very difficult task.

When protons or other common particles smash into each other, they don't often produce Higgs bosons, even when smashed together with enormous amounts of energy. The reason for this is that the particles inside protons—up quarks, down quarks, and gluons—each have very little mass. This is problematic for Higgs hunting. Whereas photons only interact with particles with electric charge and gluons only interact with particles with color, the Higgs boson only interacts with particles with mass. With little mass, the chance of creating a Higgs boson is very small.

All hope is not lost, however. Light particles such as the quarks inside of protons can still occasionally produce Higgs bosons indirectly, through intermediate particles. Consider, for example, a process such as the one depicted in the figure on the following page.

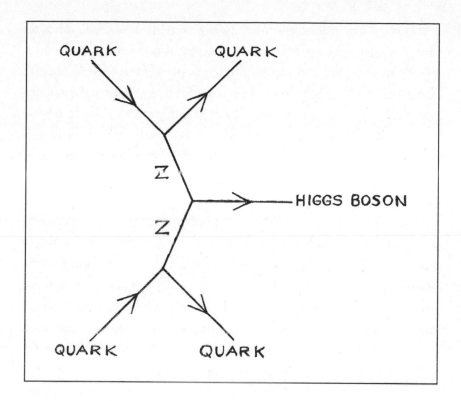

Here, two quarks—one from each of the colliding protons—come together, and each radiates or emits a Z boson. Unlike the quarks themselves, the Z bosons are very heavy—many thousands of times heavier than up or down quarks. When the heavy Z bosons come together, these heavy particles are able to transform into a mass-ophilic Higgs boson. As a result, quarks go into the interaction and quarks plus a Higgs boson come out.

Although these kinds of processes are possible, they are unfortunately very rare. Higgs bosons are expected to appear in about one out of a trillion of the collisions at the LHC. This means that if one hopes to detect the presence of the Higgs, she or he will have to look through trillions of events and determine whether they do or do not include a Higgs boson. It is like looking for a needle in a haystack.

Or is it? Consider for a moment how well this saying applies to searches for the Higgs at the LHC. A typical needle is perhaps 4 or so centimeters long and a few millimeters thick, for a total volume of about 0.2 cubic centimeters. The volume of a typical haystack, in contrast, is probably something like 20 cubic meters, leading to a needle-to-hackstack ratio of around 1 to 100 million. So looking for the one-in-a-trillion Higgs event at the LHC isn't like looking for a needle in a haystack, it is like looking for a single needle in ten thousand haystacks!

But as big a challenge as this is, the scientists at the LHC are confident that they are up to it. It might take the LHC some time—perhaps a couple of years or more once the experiment is up and running—but when all is said and done, the Higgs boson will be within reach. As Jos Engelen, the chief scientific officer at the CERN laboratory where the LHC is located, has said, "We are going to make that particle, or we are going to show that it doesn't exist." That about sums it up.

For many particle physicists, it is hard to imagine the Higgs boson or something like it *not* existing in nature. The Standard Model has been wildly successful, having been confirmed with high precision by many different experiments. It is not easy to account for the success of this theory without the last missing piece of the Standard Model— the Higgs boson—existing as well. That said, until the Higgs boson is in fact found, we will not know for certain. If it does not exist, it will be a huge surprise—and a colossal failure of theoretical physics. As the physicist John Ellis said, "If you see nothing, in some sense then we theorists have been talking rubbish for the last 35 years." I can't say that I disagree with this sentiment. But, then again, I am pretty damn sure that the LHC will discover the Higgs boson.

WHAT I AM ABOUT to say might come as a surprise to some of you, but I am going to say it anyway. I find the search for the Higgs boson a little . . . well, boring. Now, don't get me wrong; it's not that I think the Higgs itself is boring. On the contrary, the way that it transforms our universe by giving particles their mass is utterly fascinating. And it is certainly not that I think the search for the Higgs boson is unimportant. To finally nail down this particle's existence once and for all and to begin to measure its properties in detail are essential tasks, and are rightfully among the top priorities of the LHC. What I find a little less than thrilling about the search for the Higgs boson is that even before the LHC's search begins, I'm already pretty sure that this particle exists. If I had to set the Las Vegas odds for the Higgs boson existing in nature, I would probably pick something in the ballpark of 99 to 1 in its favor. Although not all physicists feel this way, many of us do—especially among theoretical particle physicists. For us, waiting for the Higgs to be discovered is a little like sitting down to watch the best team in the NFL play against the worst team—you are pretty confident you already know how it is going to turn out. To get really excited, I want something to be discovered that is less predictable, and is different from anything else we have ever found before in nature—something that represents an entirely new kind of principle behind our universe's laws. For these reasons, I am completely and absolutely thrilled by the prospect of witnessing the discovery of supersymmetry.

The superpartners predicted by supersymmetry, along with the Higgs boson, are the particles most sought after by the scientists of the LHC. But whereas we have a fairly good idea of how the Higgs boson will appear in this experiment, supersymmetry could take a vast array of forms, and could manifest at the LHC in many different ways. After all, supersymmetry isn't just a single particle to search for, but lots of particles—squarks, gluinos, charginos, sleptons, neutralinos, and so on. Depending on the various properties of each of these different superpartner particles, they could be created and appear in many different ways. We do not yet know what form supersymmetry takes in nature. Soon, however, the LHC will almost certainly tell us a great deal about it.

The superpartners that are likely to be produced in the greatest numbers at the LHC are those which interact through the strong force—squarks and gluinos. Such particles don't survive for long inside the detectors of the LHC, however. In fact, they decay almost instantly. Well before any squarks or gluinos reach the layers of electronics that make up the detectors, they will disintegrate into other kinds of particles. To recognize that squarks or gluinos were created, the detectors are going to have to identify the particles that are produced in these decays.

This is the same approach that had been taken in the 1980s when the SPS scientists were looking for—and briefly thought they had found—supersymmetry in the form of monojet and dijet events (see chapter 8). The simplest way a squark or gluino can decay is into a quark or gluon—which is seen at a collider as a jet of energetic particles—along with a neutralino that disappears undetected, noticed only as "missing energy." This jet plus missing energy signal is what SPS called a monojet. Slightly more complicated squark or gluino decays can lead to two jets and missing energy—dijets.

The same idea essentially applies to searches for supersymmetry at the LHC, but with a few more complications. At the LHC, a typical process featuring gluinos and squarks might look something like the figure on the next page.

Gadzooks, what a mess! Here a quark and a gluon, through the exchange of a squark, collide and become transformed into a gluino and a squark. These particles each decay almost instantly into a squark and a quark or gluon. The squarks, in turn, decay into more quarks, along with either a neutralino or a chargino. The chargino goes on to decay into a neutralino, along with a muon and a neutrino. When all is said and done, four jets fly out of the detector, along with a track from the muon, and a lot of missing energy

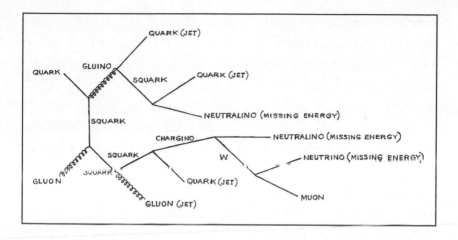

corresponding to the undetected neutrino and neutralinos. And this is only one example of many supersymmetric processes that could take place at the LHC. Squarks and gluinos could potentially decay in a large number of ways, leading to a wide variety of observational signals. Ultimately, however, every supersymmetric particle that is produced ends up as missing energy in the form of neutralinos, along with a combination of jets and electrons, muons, or taus. Although they can be very complicated, if enough events like these are seen at the LHC, it will be the first direct evidence for supersymmetry.

SEARCHING FOR THE HIGGS boson at the LHC is likely to be something of an arduous task. Very, very few of the LHC's collisions lead to the creation of a Higgs, so in order to produce and observe enough of these particles to identify them conclusively, the accelerator will likely have to run for years. Exactly how long this takes depends on how heavy the Higgs boson is. In any case, this particle is not likely to be seen very early on at the LHC.

Supersymmetric particles, however, are very different from Higgs bosons. If they in fact exist, then the strongly interacting squarks and gluinos will be created thousands of times more frequently than Higgs bosons. Once the accelerator is up and running and all of the calibrations and other tests of the detectors have been completed, evidence for supersymmetry should appear in the data of the LHC almost immediately. If the squarks and gluinos are not too heavy, then even a few weeks worth of data could be enough to discover the first supersymmetric particles. The LHC is a fantastic machine for discovering supersymmetry—especially to impatient physicists like me.

But the quest for supersymmetry does not end so quickly. As new particles begin to rear their heads, the task of discovery becomes transformed into the task of understanding what these particles are, and what their existence means for the overall structure of our Universe and its laws. To begin with, an important question that will need to be answered is whether any newly discovered particles are in fact supersymmetric particles. In other words, has supersymmetry really been discovered, or might the new particles be associated with some other kind of new physics? I've heard more than one cynical physicist make the prediction that "supersymmetry will certainly be discovered at the LHC . . . regardless of whether or not it actually exists." This sentiment stems from what I like to call the "Christopher Columbus syndrome." When Columbus set out on his famous voyage in 1492, he intended to discover a fast western passage to Asia. When he found new lands in the Caribbean Sea, he was willing and eager to call them Asia. Similarly, when the first new particles are discovered at the LHC, there is the danger that they will be called supersymmetric particles . . . even if they are not.

Physicists have proposed many theories over the years that could potentially be misidentified as supersymmetry at the LHC. In some of these models, for example, particles traveling through extra dimensions of space beyond the three we experience appear to us as very massive versions of ordinary particles, called Kaluza-Klein states. The potential for confusion comes from the fact that a Kaluza-Klein particle—a particle traveling through extra dimensions of space—will look a lot like a superpartner particle to the detectors of the LHC. There are also other theories—called little Higgs models—that predict new kinds of particles that might be confused with superpartners. To determine conclusively which theory—whether supersymmetry or otherwise   the LHC is observing, many different measurements are going to need to be scrutinized. A great deal of input from theoretical physicists will also be required to devise ways of distinguishing between the various theories.

One way that physicists have begun to prepare for these challenges is by holding a series of meetings known as the "LHC Olympics." At these workshops—the first of which was held in 2005 at the LHC's site in Geneva—groups of physicists compete to best analyze simulated, or "fake," LHC data. A few months before each meeting, fake sets of data designed to mimic what the LHC could hypothetically record when it begins its operation are given to the competing physicists. Over the next months, the physicists scrutinize the sets of "data" and try to determine what kind of physics model—supersymmetry or otherwise—has been written into the data by the conference organizers. After each group presents their interpretations of the data, they are told what "true" theory had been contained within it.

The point of the LHC Olympics meetings is not for one group of physicists to prove they are better at data analysis than another, but instead for the various groups to learn from their own and each others' mistakes. No champions are named and no medals are given. We all know that a true spirit of cooperation is going to be needed if we are going to learn as much as possible from the LHC. This is to be, after all, the experiment of a generation.

> Beauty is truth, truth beauty. That is all ye
> know on Earth, and all ye need to know.
> —*John Keats*

> The pursuit of truth and beauty is a sphere of
> activity in which we are all permitted to remain
> children all our lives.
> —*Albert Einstein*

|||||||||| **11** ||||||||||

In Search of Beauty and Truth

Walk through the halls of almost any major university's physics department and you will find signs—journal articles, scribblings on chalkboards, posters, and books—of supersymmetry's impact on the scientific community. Essentially every particle physicist in the world is familiar with the idea of supersymmetry. Thousands of us conduct research on this subject. We give lectures and write books about it. Sometimes we talk about it at each other's homes or over beers at the local pub—often to the irritation of our nonphysicist spouses and friends. The particle physicists of the world are collectively fascinated with supersymmetry. And all of this attention is despite the fact that there is no experimental evidence of its existence. As of now, there is still no proof that nature is supersymmetric. So why are we so fascinated?

There is a long list of reasons for the widespread and overwhelming interest in supersymmetry. As I've described earlier, it provides a simple way to prevent the mass of the Higgs boson from growing out of control—a solution to the so-called hierarchy problem. It also enables the three forces of the Standard Model to evolve to a single strength—an essential feature of grand unified theories. Additionally, a viable candidate for the dark matter observed in our universe—the neutralino—is present in supersymmetric models. The problems that supersymmetry seems effortlessly to solve are many.

But for many particle physicists, supersymmetry's appeal does not begin and end with the solutions it offers to these problems. For one thing, there are potentially other ways in which these problems might be solved. There are theories other than supersymmetry that might be able to alleviate the hierarchy problem. There are also other types of hypothetical particles that if they were to exist could lead to the unification of the forces. There are certainly other kinds of particles that could constitute the dark matter. So why are so many particle physicists so fixated on supersymmetry instead of these other theories? Because supersymmetry provides a single solution to all of these problems in a way that is undeniably simpler, more elegant, and more beautiful than any other theory to have been proposed. If our world is supersymmetric, all of the puzzle's pieces fit together nicely. The more that we study supersymmetry, the more compelling the theory becomes.

THE BIRTH OF NEW ideas in science can take many forms and doesn't seem to follow any simple or easily recognizable pattern. Isaac Newton formulated his theory of gravitation while working largely alone over a period of more than two decades. In contrast, the foundational laws of quantum physics were developed over only a few brief years with important contributions from many physicists—Heisenberg, Dirac, Schrödinger, Bohr, Born, and others.

On the one hand, Charles Darwin wrote his six-hundred-page *On the Origin of Species* slowly, over more than two decades. On the other, Albert Einstein wrote and published five short papers in 1905 that together proposed the special theory of relativity, demonstrated that matter consists of atoms, and introduced the first elements of quantum physics. There is little in common between how these different ideas came to be. Discovery can come in a vast array of forms.

In my own reading of the history of science, I have always found the personal motivations of those who have invented and introduced new scientific ideas to be especially interesting. These motivations depend very much on the individual scientist, and can vary wildly—just like the discoveries themselves. The most common of these motivations are largely pragmatic—efforts to find a theory that better describes what is observed in nature, for example. In these cases, there is little in the way of personal taste or philosophy. This is simply the embodiment of empiricism and the scientific method. To the pragmatic scientist, the prospect of finding a better description of his or her world is all the motivation that is needed.

But there have also been a great many scientific discoveries that were driven by far-less-practical motivations. Einstein is, perhaps, the quintessential example of such a scientist. Although he surely wanted to find a way to describe the world more accurately and completely, he also had very strong prejudices about how he thought the world *should be*. He often let these preconceived notions guide him. While developing the theory of relativity, for example, Einstein was not strongly motivated by a desire to resolve anomalous or unexplained data.[1] Instead, he was driven largely by his desire to

---

[1] While Einstein was developing his special theory of relativity, a device called the Michelson interferometer had already been used to show that the speed of light is the same in different frames of reference—in other words, observers moving at

formulate a theory that was simple and elegant, and would resolve theoretical—and even philosophical—questions. In contrast, he seemed far less concerned with the results of experiments. Shortly after his general theory of relativity was observationally confirmed in 1919, he was asked what he would have done if the expedition had found another result, and thus failed to support his theory. Einstein replied, "I would have had to pity our Lord. The theory is correct all the same." Although this remark was probably intended to be tongue in cheek, it is true that Einstein was thoroughly convinced that his theory was correct—well before it had been empirically tested.

Scientists such as Einstein can be profoundly motivated by their own notions of how the laws of our universe should be. With his famous words "I shall never believe that God plays dice with the world," Einstein argued that the laws of nature must be deterministic, and not merely probabilistic.[2] He did not ground this argument in any kind of observation or experiment. It was nothing more than a very strongly held philosophical preference. To Einstein, a probabilistic world was an ugly prospect. Up until his death, he insisted that the universe must be mechanistic and deterministic, moving forward and changing in a predictable way. He was unwilling to accept the philosophical implications of what he perceived as an ugly blemish on nature's blueprint.

Many physicists today are also driven—at least to some extent—by their ideas about how the world should be. Theories that are elegant and simple, and at the same time powerful and universal, can possess great appeal to scientists. Supersymmetry is a fantastic example

---

different speeds all see light moving at the same rate. Although the result of this experiment could have provided a great deal of motivation for relativity, Einstein didn't learn about these findings until after he had published his theory.

[2] Although Einstein often invoked the word *God* to communicate how he felt about the universe and its laws, he was actually an outspoken atheist. Unfortunately, this fact is often misrepresented—deliberately and otherwise.

of such a beautiful theory. Even in the absence of experimental evidence, we are drawn to it and expect that it somehow must be manifest in nature. But should we be wary of this siren's call? How much confidence should we place in beauty's role as a guide? Must the universe be beautiful in this way? The answers to these questions are not at all clear. Although Einstein's conception of beauty in the laws of nature often helped to guide him to new and important insights, it also sometimes misled him. For purely philosophical reasons, he died rejecting many of the basic tenets of quantum physics. Nearly all physicists today see Einstein's stubborn adherence to determinism as an embarrassing mistake. Beauty, it seems, can be a powerful guide—but at times, it can also be a deceptive one.

JOHANNES KEPLER WAS ONE of the greatest and most influential astronomers of all time. He was the first person to realize that the orbits of planets follow ellipses around the Sun, rather than circles, and discovered the three laws of planetary motion that bear his name. It was by building upon these discoveries that Isaac Newton was later able to formulate his theory of gravitation. In many ways, the mathematical foundations of physics began with the work of Johannes Kepler. Not all of his theories were correct, however. Much like Einstein, Kepler is a spectacular example of a scientist who was misled by his philosophical preconceptions about how the world should be.

Long before Kepler had discovered his laws of planetary motion, he wrote and published his first astronomical work, *Mysterium Cosmographicum*. This book contained within its pages one of the grandest and most elegant and beautiful theories ever to describe the structure of our Universe and its laws. The cosmology laid out by Kepler in this book is undeniably compelling . . . and undeniably wrong.

The story behind Kepler's *Mysterium Cosmographicum,* as it is often told, began one day in 1595. While delivering a lecture on astronomy, he had a sudden epiphany. In that moment, Kepler recognized a pattern in the orbits of the six planets known at the time. It seemed that he had discovered an elegant and simple geometric design hidden within the planets' locations.

For thousands of years, geometers have considered a group of five three-dimensional shapes, known as the Platonic solids, to be especially elegant and beautiful from a mathematical perspective. What Kepler noticed in his moment of epiphany was that if these five solids—the octahedron, the icosahedron, the dodecahedron,

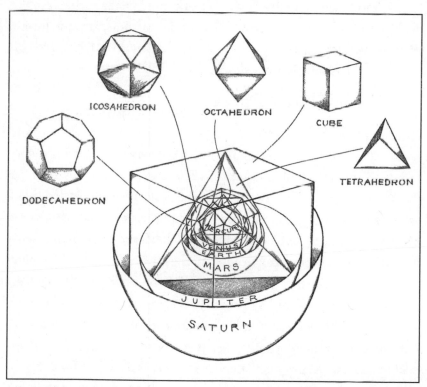

Johannes Kepler's Platonic solid model of the solar system. This is an example of a very beautiful, elegant, and completely incorrect theory.

the tetrahedron, and the cube—were encased in spheres and then put in successive layers around one another, the relative sizes of the six spheres seemed to match the observed sizes of the orbits of the six known planets. From this, Kepler became convinced that he had uncovered the design behind the planets' orbits. Furthermore, he argued that there are precisely six planets—not more nor less—because there are five Platonic solids.

That these orbits roughly coincide with the geometry of Kepler's scheme is, in actuality, nothing more than a coincidence. The Platonic solids have little or nothing to do with the dynamics of our solar system. But Kepler was convinced that he had discovered God's geometric plan for the Universe. Even after he had done his later—and much more important—work on planetary orbits, Kepler never gave up on his Platonic solid model of the solar system. He died thirty-four years later mistakenly believing the *Mysterium Cosmographicum* to be his greatest scientific contribution.

THE SOMETIME PRECARIOUS CONNECTION between beauty and truth convinced Kepler to mistakenly accept a wildly incorrect theory of our solar system, but also led Einstein to explore and discover the theory of relativity. The desire for mathematical beauty has led many astray, and many to great scientific accomplishments. The symmetries discovered by Paul Dirac and Murray Gell-Mann, for example, each demonstrate mathematical beauty in the highest degree. On many other occasions, such motivations have led otherwise brilliant scientists down long and fruitless paths. This leads us to ask whether, in a given instance, beauty should be seen as a reliable guide for science. In particular, we might wonder whether the beauty, elegance, and appeal of supersymmetry make it more likely to exist in nature.

In many ways, supersymmetry is similar to many of the other scientific theories that have been proposed over the years. Like

Einstein's relativity, Dirac's antimatter, or even Kepler's Platonic solid–based solar system, supersymmetry is a theory that was developed and studied largely for its mathematical properties and appeal. In other respects, however, supersymmetry is a unique case in the history of science. Supersymmetry was first developed in the early 1970s—more than three and a half decades ago—and still has not been experimentally confirmed. In contrast, Einstein's theory drew immediate attention and was tested quickly. Dirac's prediction of antimatter's existence was made only a year before the positron was discovered. By these standards, we have been waiting a very long time for supersymmetry's confirmation.

And physicists have not been idle over the decades since supersymmetry's birth. Tens of thousands of articles have been written on supersymmetry. Many major experimental efforts have been carried out in search of superpartner particles. In terms of scientific resources, much more has been invested in the theory of supersymmetry than had been in the theories of Einstein or Dirac before they were experimentally confirmed. It is probably safe to say that never before has such a colossal scientific effort been made to pursue such a speculative hypothesis.

For a scientist, there is ultimately only one way to resolve the question of whether supersymmetry is written into nature's blueprint. Regardless of the beauty of an idea or theory, we cannot say that it is true or false until we have compared its predictions to observations of our world. Experiment and observation is the final judge of any scientific hypothesis. If the LHC turns on and discovers superpartner particles, then supersymmetry is a theory that is beautiful, elegant, and manifest in nature. If no such particles are ever found, then nature is not supersymmetric—at least not in the way that most particle physicists envision it to be. The incredible beauty of supersymmetry is clear. The truth of supersymmetry is not. With the coming of the LHC, however, this is about to change.

One way or the other, we will soon know whether our world is supersymmetric.

The philosopher Descartes said, "Truths are more likely to have been discovered by one man than by a nation." Although that may have been the case in his day, the quest to discover supersymmetry has required the cooperation of many thousands of physicists and engineers from many of the world's nations. Supersymmetry is being searched for through an experiment of unprecedented scale. Supersymmetry is a theory of unprecedented beauty. It is a very exciting time to be a scientist.

IF I WERE WRITING the last chapter of this book a few years from now, I would wrap it up nicely by telling you how—or if—supersymmetry was finally discovered. As of today, however, the LHC is only beginning its operation and thus supersymmetry has not yet been found. Although my money is on supersymmetry revealing itself within the next few years, I would be lying if I were to tell you that I was certain about how this is going to turn out. So, in lieu of me trying to predict how this story is going to end, I've decided to leave it up to you. Ever since I was seven years old, I've wanted to write a choose-your-own adventure book. Here is my chance. Turn the page and enjoy!

The LHC Choose-Your-Own-Adventure Story!

You are a third-year graduate student working toward your PhD at the University of Wisconsin. After finishing your required classes, you move to Geneva, Switzerland, to work alongside your adviser and her other students at the site of the most exciting experiment in the world—the Large Hadron Collider. One day, your adviser calls you into her office and asks you what topic you would like to base your thesis work on. "On the one hand," she says, "you could join the team that is searching for the Higgs boson. I think they are almost certain to find something sooner or later. On the other hand, you could devote your efforts to discovering supersymmetry." She also tells you about another group that is doing a more general analysis of the incoming data.

What do you tell her?

- If you say that you want to look for signs of the Higgs boson in the LHC data, turn to page 215.
- If you say that the Higgs boson is a little dull for your taste, but you would love to look for supersymmetric particles, turn to page 206.
- If you would rather remain agnostic about what the LHC is or isn't likely to find, and perform a more general analysis of the data, searching for "who-knows-what," turn to page 210.

Very nervous and very excited, you run into your adviser's office and throw the plot with the bump feature appearing in it on her desk. At first, she seems irritated by your intrusion, but after explaining the figure to her, she starts to get excited, too. "It looks like a Higgs boson that is decaying into tau leptons!"

Almost immediately, rumors start circulating around the physics community about the so-called Higgs bump. In the weeks and months to follow, you are invited to give seminars about your recent analysis. At the end of these talks, you are often asked questions such as "Do you really think that this is finally the discovery of the Higgs boson, or might this be something other than the Higgs . . . maybe even a problem with the experiment itself?"

How do you respond?

- If you try to be cautious and say, "It is too early to say for sure. We will have to wait and see what we learn after collecting more data," turn to page 217.
- If you say, "All indications are that this is the first detection of the Higgs boson," turn to page 209.

Another year passes and more data is collected at the LHC. Unfortunately, the new data does not work to your favor. Slowly, the bump gets smaller and smaller until it goes away. There was no Higgs boson signal where you had claimed it to be. It was all just the product of serendipity and chance.

Your adviser has not been impressed by your handling of this situation and tells you to stop working on the Higgs search. You end up writing your thesis on a far-less-exciting topic and never discover a new particle.

THE END

You get your wish and begin your hunt for superpartner particles almost immediately. Only a month ago, the LHC started running, and you are excited to be among the first to get a look at the data. More experienced scientists warn you, however, that data produced this early in an experiment's operation is often plagued with miscalibrations and other problems.

When you run the first data through your analysis programs, you find many more events with jets, leptons, and missing energy than the Standard Model predicts. If gluons or squarks are being produced in the collisions, however, they should decay and generate precisely these kinds of events. This looks a lot like supersymmetry!

If the detector is giving you reliable data, then it seems likely that you have just discovered supersymmetry—but it might only be a bug that needs to be flushed out. Your more conservative colleagues suggest that you should wait for these issues to be scrutinized and resolved before getting too excited. If you wait too long to announce your findings, however, physicists from another of the LHC's detectors might beat you to it.

What do you do?

- If you and your colleagues schedule a press conference to announce the discovery of supersymmetry, turn to page 216.
- If you decide to wait and see how things develop, turn to page 212.

You send off an e-mail describing your signal to a theorist you met once at a conference. Only a few minutes later, your phone rings and he is on the line. "This is exactly what I have been predicting for years!" he exclaims. You calm him down long enough to hear him explain that the signal you have discovered is not supersymmetry, but instead looks like particles *traveling in extra dimensions of space!*

If this doesn't win you the Nobel Prize, nothing will.

Congratulations, you have discovered something even more spectacular than what you were looking for!

THE END

The more you look at the data the more you are convinced that the signal is real—regardless of how strange it is.
What to do?

- If you decide to ask a theoretical physicist what this strange signal might be, turn to page 207.
- If you ask your adviser what her opinion is, turn to page 213.

Pretty soon, physicists all around the world are talking about the detection of the Higgs boson that you have announced. The media—newspapers, radio, and television—have all been covering it. You and your colleagues are very excited! Then one day, physicists at one of the other LHC detectors announce that they have looked at their data and don't see any such bump. It could be that their detector or their analysis methods aren't as good at detecting this signal as yours are, but some of your colleagues are worried that your bump might not be a detection of the Higgs after all. Is it possible that you made a mistake? You begin to feel nervous.

What do you do?

- If you say, "In light of these new results, we will need more data before we know whether this is or is not a Higgs boson signal," turn to page 217.
- If you stand by your result and say, "I don't care what the other detector says, we have detected the Higgs!" turn to page 205.

You decide that you are the kind of scientist who says, "Who knows what this incredible machine is going to find? After all, why look for one kind of new particle when you could look for everything!"

After the LHC has been running for about a year, you are given the first set of data to scrutinize. You begin to look for departures from the predictions of the Standard Model. You look for events with lots of missing energy or with too many jets, and check to see if any unexpected bumps appear. After a while, something weird begins to stand out. There are huge numbers of very strange events scattered throughout the data.

What is going on?

- If you think you must be analyzing the data incorrectly, and keep working on it yourself, turn to page 208.
- If you decide to show these strange results to your adviser, turn to page 211.

Your adviser looks at the data and immediately says, "This doesn't make any sense at all. Come back when you have analyzed it properly." You walk away embarrassed, wondering if you will ever graduate.

Sadly, you never discover any new particles.

THE END

Your better judgment compels you to wait and see how things develop before going public with the possible detection of supersymmetry. You begin to double- and triple-check each piece of your analysis and, sure enough, after a few months of careful scrutiny, you discover a major bug in one of your computer programs. Once corrected, your signal changes considerably, and you are no longer sure whether you are or are not seeing evidence of superpartner particles. One thing that you are sure of, however, is that the signal doesn't look like the ordinary particles of the Standard Model.

What do you do?

- If you decide to ask a theoretical physicist you know what this strange signal might be, turn to page 207.
- If you decide to work more on your analysis yourself—after all, you might have just made a mistake—turn to page 214.

" I know what this looks like!" exclaims your adviser. "It seems that you've found a new particle that decays into tau leptons—maybe even a Higgs boson." She tells you to plot the data in a different way and, when you do, a "bump" feature appears.

Turn to page 217.

Months have passed, and you have checked every step of your analysis more times than you can count. Although some parts of the analysis have improved in the process, the results haven't changed much. As time goes on, you and your colleagues become more and more confident that you have actually discovered super-symmetry! Your adviser organizes a press conference to make the announcement.

Turn to page 216.

You join the group of physicists conducting the Higgs boson search and immediately go to work. As time goes on, you and your colleagues increasingly understand the incoming data, and find ever more effective ways of Higgs hunting. As the LHC finishes its third year of operation, you continue to scour through the data collected so far, looking for any signs that Higgs bosons are being produced. As months and more months go by, you begin to wonder whether evidence of this elusive particle will ever materialize. Then one day, just when you are about to lose heart, you notice an unexpected "bump" in some of the data. After a quick calculation, you estimate that such a bump is 99 percent likely to be the result of some kind of new process or particle.

What do you do?

- If you think the bump is real—maybe even be the Higgs boson you are looking for—and rush to tell your adviser and other colleagues immediately, turn to page 204.
- If you decide to wait and see whether the bump either goes away or gets more pronounced with more data, turn to page 217.

The day after your press conference, the front page of the *New York Times* reads, "Scientists Discover New Fundamental Symmetry of Nature." The excitement and attention are almost too much to take. The scientific community fills its champagne glasses and basks in your historic discovery.

As time goes on, hundreds of scientists scrutinize the data and, although some minor improvements are made in the analysis, the superpartner particles you discovered are confirmed to be real.

Start packing your bags for a visit to the Nobel committee in Stockholm. You have discovered supersymmetry!

THE END

You and your colleagues decide to play it safe and temper your public statements until you are sure about the validity of the bump feature. Over the next two years, you watch as more and more data is collected at the LHC. Slowly, the bump becomes more and more pronounced. You have scrutinized every single one of the potential backgrounds, and have become very confident that none of them are faking the observed feature. You talk to a number of leading theoretical physicists, who are each increasingly convinced that the signal matches that of a Higgs boson—in particular, the kind of Higgs that is found in supersymmetric models. After all of the scrutiny, everyone agrees that you have a discovery on your hands.

Congratulations! You have discovered a Higgs boson!

THE END

Index

Page numbers in *italics* refer to illustrations and tables.